Korea and
the Knowledge-based
Economy

Making the Transition

THE INTERNATIONAL BANK FOR RECONSTRUCTION AND DEVELOPMENT/THE WORLD BANK
ORGANISATION FOR ECONOMIC CO-OPERATION AND DEVELOPMENT

ORGANISATION FOR ECONOMIC CO-OPERATION AND DEVELOPMENT

Pursuant to Article 1 of the Convention signed in Paris on 14th December 1960, and which came into force on 30th September 1961, the Organisation for Economic Co-operation and Development (OECD) shall promote policies designed:

- to achieve the highest sustainable economic growth and employment and a rising standard of living in Member countries, while maintaining financial stability, and thus to contribute to the development of the world economy;
- to contribute to sound economic expansion in Member as well as non-member countries in the process of economic development; and
- to contribute to the expansion of world trade on a multilateral, non-discriminatory basis in accordance with international obligations.

The original Member countries of the OECD are Austria, Belgium, Canada, Denmark, France, Germany, Greece, Iceland, Ireland, Italy, Luxembourg, the Netherlands, Norway, Portugal, Spain, Sweden, Switzerland, Turkey, the United Kingdom and the United States. The following countries became Members subsequently through accession at the dates indicated hereafter: Japan (28th April 1964), Finland (28th January 1969), Australia (7th June 1971), New Zealand (29th May 1973), Mexico (18th May 1994), the Czech Republic (21st December 1995), Hungary (7th May 1996), Poland (22nd November 1996) and Korea (12th December 1996). The Commission of the European Communities takes part in the work of the OECD (Article 13 of the OECD Convention).

WORLD BANK INSTITUTE

The World Bank Institute (WBI) provides training and other learning activities that support the World Bank's mission to reduce poverty and improve living standards in the developing world. WBI's programs help build the capacity of World Bank borrowers, staff, and other partners in the skills and knowledge that are critical to economic and social development.

WBI is located at World Bank headquarters in Washington, DC. Many of its activities are held in member countries in cooperation with regional and national development agencies and education and training institutions. The Institute's distance education unit conducts interactive courses via satellite links worldwide. While most of WBI's work is conducted in English, it also operates in Arabic, Chinese, French, Portuguese, Russian, and Spanish (*www.worldbank.org/wbi*).

Library of Congress Cataloging-in-Publication Data has been applied for.

Foreword

Knowledge is fast becoming a key factor in economic and social development worldwide. Rapid innovations in science, communications and computing technologies are opening up new opportunities for countries to harness knowledge and participate more fully in the global economy. Developing countries that successfully make the transition to the knowledge-based economy will have unprecedented possibilities to become more competitive in world markets and to participate in the global information society. New technologies can also extend the benefits of knowledge to all segments of society and help countries close the gap in living standards among their citizens.

This book defines a knowledge-based economy as one where knowledge is created, acquired, transmitted and used effectively by enterprises, organizations, individuals and communities. It does not focus narrowly on high-technology industries or on information and communications technologies, but rather presents a framework for analyzing a range of policy options in education, information infrastructure and innovation systems that can help usher in the knowledge economy. It also makes the case for better co-ordination among the government, the private sector and civil society to enhance competitiveness and advance economic and social development.

The book, which is based on a joint study by the World Bank and the Organisation for Economic Co-operation and Development, breaks new ground in its attempt to develop a comprehensive set of national policy responses to the knowledge revolution. The study focuses on Korea, a country with limited natural resources which has developed mainly through an outward-oriented, industry-led strategy based on large firms and economies of scale. Today, however, this industrial paradigm is being challenged by the rapid rise of knowledge as the principal driver of competitiveness.

We believe that *Korea and the Knowledge-based Economy* provides useful lessons for other countries as they move away from old paradigms toward new models of development.

Vinod Thomas	Herwig Schlögl
Vice-President of the World Bank Institute	Deputy Secretary-General of the OECD

Acknowledgements

This study was prepared at the request of the Ministry of Finance and Economy of the Government of Korea as an input to its strategy for becoming an advanced knowledge-based economy. The timing of the study was appropriate as the government was in the final stages of preparation of its three-year Master Plan, following the announcement by President Kim in January 2000. It is the result of a collaborative effort between the World Bank and the OECD, with the bulk of the funding being provided by the World Bank's Korea country program budget.

During the preparation of this study over the past year, several missions made up of World Bank and OECD staff visited Korea and held discussions with various government counterparts, representatives from think tanks, the private sector and civil society, to obtain a better understanding of the issues confronting Korea in its transition to a knowledge-based economy. During these meetings, the team gained valuable insights on the process, timeline and details of the Korean vision and strategy.

Carl Dahlman undertook the overall co-ordination of the report and headed the World Bank team, which included Saha Dhevan Meyanathan, Anuja Adhar Utz, Jean-Eric Aubert (on secondment from the OECD), Moon Kyu Bang, Xiaonan Cao, Swati Ghosh, Denis Gilhooly, Charles Kenny and Eul Yong Park (consultant).

The OECD team was headed by Thomas Andersson, and included Peter Avery, Zhang Gang, Jean Guinet, Abrar Hasan, Daniel Malkin and Joonghae Suh (consultant).

Valuable contributions and comments were received from World Bank colleagues, including Charles Abelmann, Sri-Ram Aiyer, Jacqueline Baptist, Gillian Brown, Bruce Harris, Emma Hooper, Masahiro Kawai, Homi Kharas, Eun Jeong Kim, Bruno Laporte, KatherineMarshall, John Middleton, Richard Newfarmer, Zia Qureshi, Christopher Thomas, and Shahid Yusuf, as well as from Randall Jones and Dimitri Ypsilanti of the OECD. Research assistance was provided by Sean White and Zhihua Zeng. Administrative assistance was provided by Megan Breece.

We would like to express our appreciation to our Korean counterparts for their many insightful comments, notably those received, at the time of writing, from Mr. Kun Kyung Lee, Deputy Minister, Ministry of Finance and Economy, Mr. Yong Lin Moon, Minister of Education, Mr. Nyum Jin, Minister of Planning and Budget, Mr. Oh Seok Hyun, Managing Director of the National Economic Advisory Council to the President of Korea, Mr. Jin Soon Lee, President, Korea Development Institute, Mr. Jungho Yoo, Vice President, Korea Development Institute, and Mr. Cheonsik Woo, Korea Development Institute.

Table of Contents

List of Tables

List of Figures

Annex Figures

Currency Equivalents

As of 21 June 2000

Currency unit = Won (KRW)

USD 1.00 = KRW 1 119.20

KRW 1.00 = USD 0.000893

Abbreviations

ADSL	Asymmetric Digital Subscriber Line	KSMBA	Korea Small and Medium Business Administration
ANVAR	Agence nationale de valorisation de la recherche (French Innovation Agency)	KT	Korea Telecom
CDMA	Code Division Multiple Access	LCD	Liquid Crystal Display
Chaebol	Conglomerate	MIC	Ministry of Information and Communications
CNRS	Centre national de la recherche scientifique(National Center for Scientific Research)	MNC	Multinational Corporation
DDSA	Door-to-door Sales Act	MOCIE	Ministry of Commerce, Industry and Energy
EPCC	Economic Policy Co-ordinating Committee	MOE	Ministry of Education
ERC	Engineering Research Center	MOFE	Ministry of Finance and Economy
FDI	Foreign Direct Investment	MPB	Ministry of Planning and Budget
FSC	Financial Supervisory Commission	NEAC	National Economic Advisory Council
FTA	Fair Trade Act	NII	National Information Infrastructure
GDI	Gender-related Development Index	NIS	National Innovation System
GDP	Gross Domestic Product	NPL	Non-Performing Loan
GEM	Gender Empowerment Measure	NRDP	National R&D Program
GNP	Gross National Product	OECD	Organization for Economic Co-operation and Development
GRI	Government Research Institute	PC	Personal Computer
GSM	Global System for Mobile Communications	PCER	Presidential Commission on Education Reform
ICT	Information and Communication Technology	POSTECH	Pohang Science and Technology Institute
IMD	International Institute for Management Development	PPP	Purchasing Power Parity
		R&D	Research and Development
IMTS-2000	International Mobile Telecommunication Standard 2000	RRC	Regional Research Center
ISDN	Integrated Services Digital Network	SAT	Scholastic Achievement Test
ISO	International Standards Organization	SCI	Science Citation Index
KAIST	Korean Advanced Institute for Science and Technology	SME	Small and Medium Enterprise
KAMCO	Korea Asset Management Corporation	SNU	Seoul National University
KBE	Knowledge-based Economy	SRC	Science Research Center
KCC	Korean Communications Commission	STEPI	Science and Technology Policy Institute
KDI	Korea Development Institute	STI	Science Technology and Innovation
KEDI	Korea Education Development Institute	STSP	Specialized Telecom Service Provider
KEPCO	Korea Electric Power Company	TFP	Total Factor Productivity
KFTC	Korea Fair Trade Commission	TMA	Trademark Act
KIET	Korea Institute for Industrial Economics and Trade	TRIPS	Trade-Related Aspects of Intellectual Property Rights
KINITI	Korea Institute of Industry and Technology Information	UI	Unemployment Insurance
KIS	Korea's Innovation System	UNCPC	United Nations Central Products Classification Scheme
KITA	Korea Industrial Technology Association	VoIP	Voice over Internet Protocol
KOSDAQ	Korea Securities Dealers Automated Quotation (a NASDAQ-like stock exchange)	WIPO	World Intellectual Property Organization
KOTRA	Korea Trade Promotion Corporation	WTO	World Trade Organization
KSE	Korea Stock Exchange		

Executive Summary

This executive summary provides an overview of the study. The first section summarizes the challenge of the knowledge revolution to Korea's development strategy. Section B describes the analytical and policy framework for a knowledge-based economy used in this report. Section C highlights the main issues in the four key areas of the framework: the economic incentive and institutional regime; education, training and human resource management; information infrastructure; and the innovation system. Section D discusses industry-related issues and Section E makes some observations on the implementation of these reforms in Korea, based on the experiences of other countries. Details on the specific findings and recommendations are presented in the matrix in Section F.*

A. The knowledge revolution: the challenge to Korea's development strategy

Korea's achievements. Korea has achieved one of the fastest rates of economic development of any country in the world. Between 1966 and 1996, its per capita income grew by an average of 6.8% per annum,** and it became an OECD Member in 1996. Towards the end of 1997, however, Korea experienced its worst economic crisis since the Korean War. Nonetheless, Korea made a remarkable recovery from the crisis and its economy grew at 10.7% in 1999. The government expects some 8% growth in 2000 and around 6% in the following few years. However, this performance may not be sustainable.

Irrespective of the crisis, Korea faces a difficult and competitive global environment. Its wages have risen and it is experiencing increasing competition from lower-wage countries in East Asia. Although its manufactured exports have been expanding rapidly, it is being squeezed between the developed OECD countries at the higher end, and China and other East Asian developing countries at the lower end. As a result of these developments and the increasing importance of knowledge, Korea is confronted with the unprecedented challenge of transforming itself into a knowledge-based economy.

The global knowledge revolution. Increases in scientific understanding and very rapid advances in information and communication technologies (ICTs) mean that knowledge and information have become key to competitiveness. Technical progress and reductions in transportation and communications costs are leading to a more interdependent and competitive world. Investors are increasingly seeking first-mover advantages, new products and services in response to customers' diverse and rapidly changing demands, speed to the market, and first-rate access to customers and sources of information. The rapid development of ICTs and the Internet is exposing inefficiencies in the functioning of markets, firms and institutions, putting downward pressure on prices and accelerating the need to restructure and adapt to changing conditions. ICTs are also improving the efficiency of interaction among government agencies and the delivery of government services, as well as facilitating consultation with the public. This is bringing a new potential for the re-organisation of administrative and political institutions and for dramatic reductions in the cost of delivering services. In addition, new ways and means of networking are altering social patterns of work, shopping, education, leisure and communications. These changes have given rise to three overarching challenges for Korea's future development strategy.

---

* The matrix distinguishes the situation prior to the crisis in Korea, outlines ongoing reforms and identifies the outstanding issues that need to be addressed.

** In constant 1995 USD.

The first challenge is to increase overall productivity. In the past, Korea's growth was based on high capital investments and growth in labor inputs. However, the development model predicated on input-driven growth and the rapid expansion of the large *chaebol* had begun to reach its limits by the early 1990s. Although the measured rate of total factor productivity growth over the past 30 years has been respectable, it has been small relative to the rapid growth of output. Most of Korea's economic growth over the past decades has been underpinned by strong capital accumulation, made possible by a high savings rate and increases in labor input. Total factor productivity growth declined in both manufacturing and services during the 1990s. In sectors such as finance, insurance and business services, and wholesale and retail trades, total factor productivity growth actually turned negative. Growth in the future will need to be more productivity-based, requiring an increase in the efficiency of investments in physical capital and knowledge.

In Korea, investments in education, information infrastructure, and research and development (R&D) as a percentage of GDP are among the highest of the OECD economies. However, the country is not getting the full benefit of these investments because of problems with the overall economic incentive and institutional regime, as well as due to issues specific to each of these three areas. Reasons for this include:

- Inadequate conditions for generation and exploitation of knowledge and information (*e.g.* intellectual property rights and the regulatory framework with relation to IT).

- Insufficient competition, flexibility and diversity (*e.g.* the *chaebol*, the financial and education systems).

- Misallocation of investments (*e.g.* duplication of public investment in R&D, insufficient public investment in basic R&D, over-investment in an education system which is geared to simply passing exams).

These problems are serious since they cast doubt on Korea's capability to sustain its development process in the long term, despite the resources invested and recent achievements. Korea has to move to a strategy of achieving greater productivity across the board. Specific areas for concern are summarized in Section C, and developed in more detail in the attached matrix, and in the subsequent chapters of the report.

The second challenge is the need for Korea to become more internationalized in the context of an increasingly globalized and interdependent world. This includes not only opening up to trade and foreign investment, but actually striving to become more integrated in the global system. This involves:

- Developing alliances with world-class universities, fostering the exchange of professors and students, and enhancing knowledge of foreign languages and culture.

- Tapping more effectively into global knowledge systems through joint research, joint submissions to international journals, strategic alliances, inward and outward FDI, and contracting of foreign research institutes.

- More active participation and leadership in international forums and institutions, especially those that are setting the rules and practices for the new economy, such as the WTO, the World Intellectual Property Organization (WIPO), the International Standards Association (ISO), the OECD, etc.

- Ensuring harmonization with evolving international standards, including more active participation in the International Telecommunications Union, etc.

The third challenge is that, in the dynamic context of the knowledge revolution and the networked economy, the role of the Korean Government will need to be redefined. The Korean Government has played a key role in the country's rapid development and will indeed need to play an important role in the transition to the knowledge-based economy. However, it must move away from the interventionist policies of a government-led strategy based on the growth of large-scale industry, especially the *chaebol*. The Korean system has been over-regulated in the sense that the behavior of a number of actors has been constrained by "dirigiste" procedures and rigid and detailed rules. At the same time, the "rules of the game" for a competitive, transparent, equitable economy are not sufficiently developed or enforced.

Meeting the requirements of the knowledge-based economy means making markets function more effectively so that they can facilitate the constant redeployment of resources. This simply cannot be done through heavy government intervention in the economy. The key elements of the new role for government in this context are:

- Unleashing the creative power of markets.

- Providing legal and regulatory underpinnings for freer and more competitive markets – rule of law, standards for transparency and accountability, modern regulatory institutions.

- Building a modern legal infrastructure for the knowledge-based economy – intellectual property rights; cyber laws covering privacy, security, and digital transactions.

- Continuing to provide public goods (education, basic research) while addressing missing or underdeveloped markets (information networks).

- Fostering policies and institutions conducive to entrepreneurship and enterprise development (facilitating entry and exit in a number of areas, including the removal of onerous regulations hindering the start-up of new businesses especially in the service sector, promotion of high-value-added services, valuation of intangibles).

A new and important role for government will be to address the risks of the "digital divide". The ICT revolution brings with it not only opportunities, but also the risk of creating a "digital divide" between those who have access to the potential benefits of ICTs and knowledge, and those who do not. To offset the risks of a growing digital and knowledge divide, the Korean Government needs to adopt proactive policies aimed at fostering the availability of information and knowledge services, and spurring widely spread entrepreneurial activity. It should pay special attention to using ICTs and new technologies to provide opportunities for rural and poor urban communities and the disabled. In addition, the government must modernize itself in order to capture efficiency gains by improving the quality and effectiveness of public services and strengthening government information flows. Since the crisis, the Korean Government has made a very impressive start in this direction. The next step will be for it to become a facilitator of change, fostering broad participation consistent with the distributed power arrangements of the networked economy.

Korea's new vision. The Asian financial crisis has brought about a re-examination of Korea's structural problems, and a new willingness and energy to change the Korean economy and society. At the beginning of this year, the government articulated a long-term vision to transform Korea into an advanced knowledge-based nation. Goals include:

- Making Korea into one of the world's top-ten information and knowledge superpowers.

- Developing the next-generation Internet and the information superhighway by 2005.

- Promoting the use of computers by students, teachers and the military.

- Conducting radical reforms in education to arm the country for its transformation into a knowledge-based economy.

- Envisioning the dawning of an Internet society, where civil society will participate in the governance process though ICT in a democracy based on human rights.

- Closing the development divide through productive welfare and balanced regional development.

B. The framework of the knowledge-based economy

This report does not focus narrowly on high-technology industries or ICTs, but on the broader context, *i.e.* the extent to which an economy is *effectively* tapping and using the potential of the growing stock of knowledge and advances in ICT. A knowledge-based economy is defined as one where knowledge (codified and tacit) is created, acquired, transmitted and used more effectively by enterprises, organisations, individuals and communities for greater economic and social development. It calls for:

- An economic and institutional regime that provides incentives for the efficient use of existing knowledge, for the creation of new knowledge, for the dismantling of obsolete activities and for the start-up of more efficient new ones.

- An educated and entrepreneurial population that can both create and use new knowledge.

- A dynamic information infrastructure that can facilitate effective communication, dissemination and processing of information.

- An efficient innovation system comprising firms, science and research centers, universities, think tanks, consultants and other organisations that can interact and tap into the growing stock of global knowledge; assimilate and adapt it to local needs; and use it to create new knowledge and technology.

The economic incentive and institutional regime is explicitly included in this definition because it is central to the overall ability of an economy to make effective use of knowledge and is critical to the effectiveness of the three other key areas. For example, pressing on with the reforms of the economic incentive regime will induce the *chaebol* to become leaner, more competitive and more knowledge-based. It will force the exit of non-viable economic activities while providing opportunities for new entries, open up demand in the education and information infrastructure sectors, and unleash the innovativeness, creativity and entrepreneurship of the Korean people.

This framework is used in the report to take stock of where Korea currently stands and to suggest further measures to facilitate its transition to an advanced knowledge-based economy.

C. Issues in the four key areas

The challenges identified above are systemic in nature. A major task for the government is to adopt a more comprehensive and consistent approach which encompasses all the relevant policy domains. As an essential component of such an approach, reforms are necessary in the four areas discussed below, each of which contains deficiencies which have implications for Korea's ability to benefit from the knowledge-based economy.

Economic incentive and institutional regime. The main challenge is to move away from direct intervention and to foster a flexible, adaptive, market-based economy and a creative society, compatible with the knowledge-based networked economy. This means placing high priority on reforms that will enhance competition and flexibility in the economy and unleash efficiency gains and innovation. It will require measures that can bring about a fundamental upgrading of the economy's capacity for spontaneous adjustment to changing competitive pressures and opportunities and for the effective utilization and creation of knowledge. Some traditional areas will have to be deregulated, while there is a need for establishing modern regulatory oversight in certain new areas in order to strengthen markets. While the government has already started on a series of important reforms, it needs to embark on a systemic agenda for reform across a number of areas, including:

- *Product markets*: strengthen foreign and domestic competition, consumer protection and standards.

- *Financial markets*: ensure greater transparency and disclosure, strengthen corporate governance, accounting rules, prudential supervision; improve equity and venture capital markets.

- *Labor markets*: improve labor relations; make worker benefits portable and remove employment biases against women.

- *Knowledge market*: strengthen intellectual property rights, their enforcement, and the promotion and valuation of intangible assets.

- *Industrial restructuring and entrepreneurship*: put in place policies and institutions for spontaneous industrial restructuring and development, including the fostering of entrepreneurship and a more favorable performance of small and medium-sized enterprises.

- *Social issues*: mitigate the risks of the "digital divide" by upgrading social safety nets, strengthening opportunities for retraining, and extending access to education and information infrastructure to poor urban and rural communities and to the disabled.

Education, training and human resource management. During its drive for industrialization, the Korean educational system, with its emphasis on primary and secondary education and equal access to educational opportunities, served the country well and produced students of fairly high average quality. One of the main strengths of the Korean system is the high investment in education. A weakness, however, is that these investments have not been put to the most efficient use. Currently, the share of public financing in education is roughly 4.4% of GDP, lower than the OECD average of 4.9% in 1995. However, this level is supplemented by much larger investment in education by the private sector: parents contribute an additional 2.3% of GDP to tuition and other expenses at both public and private schools, from the primary to the tertiary level. In addition, lack of choice and large class size in the public sector, combined with a very strict national university admissions system, have led parents to invest another 3.2% of GDP on private tutoring for their children. An estimated additional 3.4% consists of general expenses, including the purchase of study materials, books and stationery, uniforms, transportation fees, lodging, food and other costs. As a result, Korea actually spends 13.3% of GDP on education – probably the highest share for any country at its level of development. The current environment of greater competition and rapid technical change requires that such investments be put to better use in the future, allowing people to learn and upgrade their skills so that they function more effectively in the knowledge-based economy.

The time is thus ripe for Korea to switch to a new model of education that promotes quality, creativity and lifelong learning, and that emphasizes not just formal schooling, but overall human resource development. This will entail major deregulation, decentralization and diversification of the Korean education system and enhancement of competition. The urgency of this task calls for closer and more effective co-operation between the relevant ministries, notably the Ministries of Education, Labor, Health and Welfare, Industry, Science and Technology, and Information and Communications. President Kim's recent decision to reform and reorganize the human resource development sector by setting up the post of a Deputy Prime Minister for Education is a move in the right direction. In addition, the task requires an expansion of the focus beyond the Ministry of Education to that of Human Resource Development or a Ministry of Knowledge. Some areas of the government's reform agenda that are critical for responding to the needs of the knowledge-based economy include:

- Deregulating the education system and increasing autonomy for private secondary and higher education, involving changes in curriculum and tuition, and permitting universities to set their own admission requirements, number of places, etc.

- Integrating the current formal, vocational, adult and distance education and training systems to meet the growing needs of lifelong learning.

- Reorienting the use of public and private resources to emphasize improvements in the quality of education at all levels. In addition, more scholarships should be made available for poor students to address the equity issues that will result from the growth and improvement of private education. Furthermore, special efforts should be made by the government to encourage more women to enter into higher education and into technical and scientific fields. This should be coupled with special initiatives to open up more professional employment opportunities for women.

- Introducing outcome-driven governance systems in education with clearly defined autonomy and accountability at the institutional levels and decentralization to enhance local decision making at schools and universities.

- Strengthening Korea's links to the global educational system. Universities should be encouraged to develop strategic alliances with world-class universities and encourage faculty exchange and joint courses, as has been done in Singapore. In addition, the curriculum in English and ICTs should be strengthened to facilitate global communication and international links in order to better prepare Korean students for a globalized world.

Information infrastructure. The ICT revolution will increasingly affect the efficiency and functioning of all economic and social activities. Korea has moved a long way toward liberalizing the ICT sector and is thereby capturing many of the benefits of advancing technologies. In addition, there has been a very rapid uptake of new information technologies such as mobile phones and the Internet. However, the country has a regulatory regime that may constrain continued fast development: the Ministry of Information and Communications (MIC) still tries to orchestrate much of the sector's development when, in reality, the market is moving more rapidly than the regulator can successfully anticipate. Thus, the MIC should move further toward facilitating private provision of services and addressing areas of market failure. In particular, it should:

- Liberalize the telecommunications service industry, unbundle local loop services, set up an independent regulatory agency, and open up to greater foreign investment.
- Develop a modern regulatory oversight for telecommunications that includes interconnection standards, service quality and auctioning of spectrum.
- Implement legislation on regulation of e-commerce that is harmonized with evolving international standards.

Innovation system. Korea reportedly spends more on R&D than most other OECD countries as a percentage of GDP (2.8%). However, the productivity of this effort is questionable. There are some inherent weaknesses in the Korean innovation system that need to be addressed. The government has recognized the challenge and has announced plans to increase R&D expenditures to 5% of the national budget – an ambitious target which signals the seriousness of its resolve. However, it is not so much a matter of the amount of money spent as the way in which it is spent. Key issues include:

- Encouraging greater interaction among firms, universities, government research programs and Government Research Institutes (GRIs).
- Clearly justifying the rationale for public intervention and providing subsidies in a transparent and non-discriminatory manner.
- Providing support to R&D in large companies on stricter conditions, assisting only when they would not have undertaken the concerned projects in the absence of support, and stimulating partnerships with other actors (enterprises, university and public laboratories), etc.
- Increasing the basic research effort. This measure should principally target the universities, which should receive larger resources; it implies reforming regulations and practices that discourage research activities.
- Reorienting the GRIs since their activities tend to duplicate those of industry. The GRIs have to be re-positioned to do more upstream research or to become more focused on research of collective interest (*e.g.* health, transport, etc). A larger part of their budget has to be secured in the form of institutional funding.
- Improving support to innovation in SMEs, with emphasis on effective networking and clustering, and the involvement of local authorities.
- Strongly encouraging all actors to increase their contacts with foreign counterparts, through academic and research exchanges, technological co-operation, industrial joint ventures, participation in international regulatory bodies, etc.
- Enforcing co-ordination procedures involving key ministries.
- Implementing evaluation exercises, including an international review of the country's basic research capacities.

D. Industry-related issues

In addition to the four key areas described above, a number of industry-related problems interact to reduce pressure for change and limit the ability of firms to respond to the opportunities offered by the knowledge-based economy.

For Korea to successfully master its transition to tke KBE, the government will need to take action on the following issues:

- The dominating position and limited responsiveness of the *chaebol*.

- The untapped potential of SMEs.
- Remaining barriers to investments by foreign firms.
- Impediments to high-value-added services.

The lack of incentives for firms to invest in intangible assets.

E. Implementation of the reforms

The reform strategy outlined above has to be systemic and must involve the design and implementation of measures which are consistent across different, traditionally disparate areas of policy making. While piecemeal reform of individual elements can yield some improvements, the results will not be as promising as if a series of reforms are jointly undertaken. This is particularly relevant in the knowledge-based economy where networking and horizontal interactions among policies as well as actors take on greater importance. The government has a significant role to play in making sure that all groups are well informed about the trends and forces affecting them and the need for change.

We do not underestimate the challenges – short-, medium- and long-term – that developing and effectively implementing such a systemic strategy will entail. Based on the experience of other countries that have crafted and implemented broad strategies, such as Canada, the United Kingdom, Ireland and Finland, it is clear that the development of the strategy must be undertaken in consultation with the private sector and civil society. This is particularly relevant for Korea, given the current social and political change that is taking place and the vision for the country in the 21st century enunciated by the government. The public hearings and consultations with think tanks and the private sector on national policies and reforms that have taken place so far represent an encouraging trend.

However, building consensus and buy-in from stakeholders on the desired measures requires a greater effort of dissemination, explanation and consultation with the wider public, including civil society. To illustrate, the education sector must implement major reforms – many of which have already been formulated by Korean think tanks and concerned groups. Much initial resistance is likely to come from the education agencies and from teachers, as both parties would lose some degree of power and control. It will be important to convince these groups of the benefits of the reforms.

Implementation of the strategies should also focus on whether the infrastructure to implement reforms is in place. Too often, across countries, strategies and initiatives are announced without the adequate evaluation systems and staff in place. Because of the dynamic nature of the knowledge and information revolution and the global economy, it is important to set up a monitoring and evaluation system as an integral part of the implementation process. It will also be necessary to build in provisions for adjustments to plans and actions in light of ongoing developments, including the findings of impact evaluations.

There is a tension between the need for some centralized locus of responsibility for overall coordination of the strategy and its implementation, on the one hand, and the distributed power arrangements of a networked economy, on the other. An appropriate balance has to be found in the Korean context.

These are preliminary suggestions and should be treated as points for more intensive discussions. This report reviews some of the key challenges facing a very dynamic system, seen from the perspective of two international institutions which have the advantage of a broader perspective of cross-country experiences, but which are not as knowledgeable about the specifics and details of the Korean context as the Koreans themselves. It should thus be interpreted as a first step in a broader reform process which Korea has to set up for itself, analyzing where it currently stands, where it wants to go, and how it is going to get there. Such a process has to be owned not only by the Korean Government, but by the Korean people.

F. Executive summary matrix: main findings and remaining issues

Situation up to crisis	Ongoing reforms	Remaining issues
I. Economic incentive and institutional regime		
Redefining the role of government • The government tended to be highly centralized, interventionist, and "dirigiste", especially in regard to the financial and industrial sectors.	• Since the crisis, the government has announced that it is moving more towards market-based principles. It has opted for greater openness and competition, encouraged participation of selected groups in policy discussions, and paid attention to social and safety-net issues. Greater market discipline forms the core of the new policy paradigms. Some industrial targeting remains in the telecommunications manufacturing sector.	• The government should move away from direct intervention to the provision of an incentive and regulatory framework that buttresses well-functioning markets while addressing market failures and gaps in markets, promoting public goods, and dealing with the inequities that may result. • It should foster greater entrepreneurship and spontaneous industrial restructuring and enterprise development. • The government should participate more actively in international forums and institutions, especially those that are setting rules for the new economy (WTO, WIPO, ISO, OECD, etc.).
Opening up the economy and promoting competition • A number of policy measures constrained the competitive environment, including controls on the price and quantity of credit; entry barriers and investment restrictions; tariff and non-tariff barriers; restrictions on FDI; and weak mechanisms for exit.	• The government has been reducing tariff and non-tariff barriers. • It has also strengthened antitrust actions through the Fair Trade Commission (FTC). • Deregulation efforts came with the Regulatory Reform Committee in 1998 which has a broad mandate across sectors. • In the area of insolvency and creditors' rights, the government has taken steps to improve the handling of bankruptcy cases in the court system.	• The government should continue with liberalization, especially with further deregulation in services (telecommunications, financial news, and legal services). Antitrust actions and enforcement of competition policies in general should be strengthened by giving more analytic and enforcement power to FTC. • The bankruptcy regime should be further improved, and capacity needs to be increased in the bankruptcy division of the Seoul District Court to handle its caseload.
Foreign direct investment • Historically, foreign direct investment (FDI) has been highly restricted. • Restrictions on FDI were gradually liberalized during the 1980s, including the adoption of the negative list, abolishing performance requirements, and easing friendly cross-border M&As.	• FDI restrictions have been considerably liberalized through the Foreign Investment Promotion Act of 1998. Foreign investment has increased from an average of USD 1.2 billion in 1991-95, to USD 8 billion in 1998 and USD 15.5 billion in 1999. • Restrictions on foreign land ownership were lifted in 1998.	• There should be greater liberalization of FDI into the service sector in order to increase domestic competition and provide better management models. • Accompanying reforms in corporate governance and innovation policy, including intellectual property rights, will help attract FDI that can generate positive linkages and spillovers to the economy.
Improving the soundness and efficiency of financial markets • Financial policies toward the chaebol, together with a lack of investment alternatives, channeled considerable private household savings into banks. This capital was turned into bank loans for the *chaebol* and large firms. Banks had little incentive to develop their institutional capacity to analyze credit and risks. • The non-bank sector was also compromised when the *chaebol* gained control of many intermediaries. Capital market development was hindered by pervasive explicit or implicit guaranteeing of banking instruments and weaknesses in the regulatory and institutional framework for capital markets. • Financial transparency and accountability was weak.	• Since the financial crisis, Korea has made steady progress in restructuring and reforming the financial system, reestablishing solvency, and setting up mechanisms for monitoring and supervision. • Accounting and auditing standards and practices have been upgraded and the institutions responsible for setting standards and ensuring compliance have been strengthened.	• Korea still has to resolve the overhang of non-performing loans (NPLs), adequately recapitalize banking institutions, improve the credit culture of banks, and systematize regulatory oversight of the financial sector. • It should further open the financial sector to FDI to provide more competition to domestic banks, and to benefit from global banking and management expertise. • It has yet to implement the new financial transparency and accountability standards, and modify restrictive regulations on the auditing profession.

Situation up to crisis	Ongoing reforms	Remaining issues
Corporate governance • The system of corporate governance in Korea led to massive investment and industrial expansion, resulting in over-diversification and over-leveraging, making Korea vulnerable to the financial crisis.	• Since the financial crisis, Korea has strengthened corporate transparency through improved accounting and auditing norms, strengthened minority shareholder rights, and accountability of boards of directors.	• Korea should adopt a mandatory code of corporate governance consistent with international best practice (OECD Principles of Corporate Governance), issued by the Korea Stock Exchange (KSE). • It should require listed companies to report regularly to shareholders, and remove obstacles to class action suits by shareholders against directors. • It should implement a system of secured transactions, including a national system of registration to facilitate credit to SMEs.
Venture capital • Emphasis in Korea's financial system was on collateral-based debt rather than equity. This practice presented a bias against small start-up firms in knowledge-based sectors which can incur high costs for knowledge gathering but have little physical capital that can be used as collateral. Listing in capital market was available only to well-established companies. Risk-taking is constrained by the strong cultural stigma associated with failure.	• Korea established KOSDAQ in 1996, and as of 7 March 2000, 473 firms have been listed, including 155 new venture-based firms. Value of KOSDAQ transactions is now greater than that of the KSE.	• Korea needs to strengthen regulatory oversight and information disclosure, reduce restrictions on capital funds and open full participation to foreign investors.
Enhancing flexibility in the labor market • Labor legislation, labor relations and industrial and market structures contributed to rigidity in the Korean labor market and reduced the speed with which Korea could adapt to changing competitive pressures. • There was insufficient emphasis on firm-based training and labor retraining. • There was significant employment discrimination against women.	• Since the crisis, Korea has revised its labor laws to legalize layoffs and increase the flexibility of the labor market. • The government has begun to focus on the need to provide more labor retraining.	• Korea needs to develop better industrial relations, make worker benefits fully portable, reorient training schemes to meet the demands of a more flexible economy, and ease restrictions on temporary workers. This requires greater awareness-raising and buy-in from labor. • It also needs to redress the inequality of access in job opportunities and pay for women.
Social safety nets • Korea made substantial progress in reducing poverty during the 1990s. However, with the onset of the crisis in 1998, poverty rose dramatically.	• Since the crisis, the government has begun to strengthen social safety nets and has tripled expenditures on social protection (including on public works and livelihood protection). • It began to reform its pension system in 1998, and private management and funded pensions will be given a larger role in the new system.	• Korea should build on reforms through better enforcement of labor market regulations, better targeting of programs for the poor and vulnerable, improvement of employment opportunities through retraining and enhanced labor market flexibility, and reform of the pension system.
Addressing the risks of the digital divide • The information and communication technology (ICT) revolution brings with it not only opportunities, but also the risk of creating a "digital divide" between those who have access to the potential benefits of ICTs and knowledge, and those who do not. This is likely to be even more serious in developing economies because of greater disparities in wealth and weaker social safety nets.	• As a result of the crisis, the Korean Government has become aware of the risks of increasing social inequality and, going beyond the immediate impact of the crisis, of the risk of a digital divide. • President Kim's January 2000 vision statement includes narrowing the digital divide through a "productive welfare" strategy and balanced regional development. • In the area of education, the government has plans for providing free PCs and five years of free Internet access to 50 000 students from poor families. • In the area of information infrastructure, *Cyber Korea 21* outlines initiatives such as connecting 10 400 schools; teaching civil servants, students and military personnel to use computers; building Internet plazas; and facilitating Internet PC purchases.	• The government needs to strengthen social safety nets for those who are negatively affected by economic restructuring and the associated job displacement. There is a case for broadening and further strengthening its "productive welfare" system to provide opportunities for retraining and productive employment, and foster access to new technologies, especially for the poor, women and the disabled. • It should also put in place efficient new programs providing access to new technologies, computers and the Internet for low-income families. • It should design appropriate regulatory incentives to further encourage the provision of telecommunications on a commercial basis to low-income groups in rural as well as in poor urban areas.

Situation up to crisis	Ongoing reforms	Remaining issues
Strengthening intellectual property rights and enforcement • A sizeable portion of the public had little understanding of the criminal nature of actions which amount to infringement of intellectual property rights.	• Intellectual Property Right (IPR) law was strengthened in 1998, although its enforcement is still weak. • Much of the public still considers knowledge to be a free good, thus creating a disincentive for private producers of knowledge.	• Korea needs to mount advocacy campaigns and create greater public awareness of the importance of IPRs; improve documentation and knowledge databases, and upgrade the Korea Institute of Industry and Technology Information (KINITI). • It must further improve the administration and enforcement of the laws governing IPR.

II. Education and skills

Situation up to crisis	Ongoing reforms	Remaining issues
Educational achievements and recent education reforms in Korea • Korea obtained high educational achievements compared with other OECD countries. However, the system served to constrain entrepreneurial behavior and lacked flexibility and incentives for further improvement. • The public share of education financing in Korea was 4.4% of GDP in 1998, lower than the OECD average of 4.9% in 1995. Parents spend an additional 3.2% on tutoring. • Korea had a tradition of insularity, with few strong associations with the outside world.	• The Presidential Commission on Education Reforms 1995 (PCER) made a number of recommendations, but implementation has not reached a desired level. Programs like *Brain Korea 21* have started to address this issue. • The government plans to increase the public share of education to 5% of GDP. • It has begun efforts to strengthen foreign-language education in the country. • The recent decision by President Kim to set up the position of a Deputy Prime Minister for Education to co-ordinate human resource development policies is a positive move.	• A systemic and profound reform of the current education system is necessary to spur the development of creative citizens with the necessary skills to underpin the knowledge-based economy. • There is a need for awareness campaigns to inform the public of the benefits of the proposed changes in the education system. • The educational environment should be improved to the level of the OECD countries (*e.g.* in terms of class size, public expenditures, and participation of women in science and technology). The government should reduce the financial burden on parents, while maintaining quality and equity in the education system. It should also examine the efficiency and allocation of public and private sector resources.
Deregulation and decentralization • Rigid government control over the education system included curriculum, examination system, tuition fees, and the number of students by discipline, for both public and private institutions. • Resources were used ineffectively and inefficiently and the system lacked accountability.	• PCER proposes the creation of an "autonomous school community", including the implementation of school councils and open recruitment procedures for principals and teachers. • PCER intends to establish curriculum and assessment centers for schools. • PCER aims to link financial support to universities with performance evaluation. • At the primary and secondary levels, responsibilities have been divided among the central and provincial governments and schools in designing and implementing the national curriculum. To a degree, colleges outside Seoul are allowed to determine their own tuition fees, student quotas and curriculum.	• The government should continue its deregulation efforts and encourage competition. • It should implement an outcome-driven governance system with clearly defined autonomy and accountability at the institutional level. • It should increase institutional autonomy to enhance local decision making at schools and universities. • The government should ensure equity in the system by investing in schemes to help students from poorer families. • It should establish a sound accreditation system; undertake assessments to judge the quality of teachers and programs; track and monitor quality over time; and make this information available to the public.

Situation up to crisis	Ongoing reforms	Remaining issues
Diversification		
• There was an over-emphasis on homogeneity of university types and curriculum. • There was also a geographical concentration of universities around Seoul. • All students in both private and public schools were subject to the same curriculum, and there was no streaming by ability in the general secondary school system.	• The government intends to give greater autonomy to universities in admissions (student quotas) and academic affairs. It has begun approving new specialized colleges. It has started reforming the college entrance examination to expand student choice. • *Brain Korea 21* plans to nurture leading regional universities to meet the needs of local industry by allocating KRW 350 billion from 1999 to 2005; changing the requirements for university entrance; and recruiting more professors. • The government has started to experiment with ability-based grouping at the primary level and is planning to introduce such streaming at the secondary level in 2002.	• Korea needs to promote diversity and specialization in the system through a few universities that provide comprehensive programs in a broad range of subjects, and other universities that provide specialized programs to meet different learning needs of students. It needs to ensure greater possibilities for students, ensuring self-determination and choice of subject areas, as well as mobility within the system to fulfill the goals of lifelong learning, equity and efficiency. • It should develop "double qualifying educational pathways" that can lead to the university, tertiary level vocational education or the labor market.
Relevance, quality and gender		
• There has been a shortage of training in new skills and an over-emphasis on quantitative reasoning. • Educational output was inadequately related to industrial needs. • There were insufficient linkages between the Ministry of Education and other relevant ministries. • There is also a deeply embedded gender bias against women in Korean culture and society.	• According to the 7th curricular revision that came into practice at the beginning of this year, some curriculum reform is currently underway (*e.g.* a decrease in the number of required subjects and an increase in the number of elective courses). • Proposals include reform of textbooks and teaching methods, and the use of a variety of educational technologies. • *Brain Korea 21* aims to give a greater research focus to universities by nurturing world-class graduate schools. KRW 1.4 trillion has been allocated for 1999-2005. To participate in this program, universities will be required to reform their student admission system; reduce the undergraduate student ratio in return for financial support; and develop curriculum and research programs with the world's leading universities. Performance-based promotion for professors will be also be introduced.	• Korea should integrate the curriculum to include training in new skills such as communication skills and capability to utilize ICTs, and should increase possibilities for gaining field experience. • It should enhance pedagogical training with emphasis on new knowledge and ICTs; provide incentives for teachers, including outcome- and performance-based pay schemes; and develop knowledge sharing systems. • It should implement strategies to increase the number of women participating in the economy, and change the entrenched mind-set against greater labor force participation of women in Korean society. • The government should encourage university and industry partnerships. In addition to industry providing financial support, industry representatives should be on university boards and participate in curriculum development. • It should encourage the expansion of exchange programs between Korean and foreign educational institutions; encourage entry of foreign university branches; and allow twinning arrangements. • It should strengthen the link between the Ministries of Education, of Labor, of Commerce, Industry and Energy, and of Science and Technology in order to better co-ordinate the needs of the labor market and industry with the supply of education.
Lifelong learning		
• Too much emphasis was placed on formal education compared with informal education and lifelong learning. • There was insufficient job-related training for the working-age population.	• The Education Credit Bank system has been introduced, enabling credits earned through the open educational system to be recognized as the equivalent of a formal degree. • Korea has increased its efforts to promote adult education attached to universities. • It is also focusing on ICTs (*e.g.* by wiring schools). • PCER intends to provide greater opportunities for job training for women and the elderly. The Ministry of Education will also set up a Lifelong Education Center.	• The government should establish the lifelong learning system by integrating the formal education system with distance, adult and vocational educational systems. • It should strengthen informal lifelong education programs and provide incentives and opportunities for those who have left school to reintegrate the system whenever necessary. It should fully implement the Education Credit Bank proposal. • It should assist education institutions to build up their ICT infrastructure to enable them to deliver educational training on demand.

Situation up to crisis	Ongoing reforms	Remaining issues

III. Information infrastructure

The current situation

• Until the crisis, Korea lagged significantly in terms of information-related infrastructure for an economy at its level of GNP per capita.	• Korea has undertaken very rapid investment in certain telecommunication service areas in the last few years. In 1999, mobile phones surpassed fixed lines. Internet prices are low for off-peak times and Internet use has been growing fast. Public access to the Internet is high. *Cyber Korea 21* and related initiatives (including President Kim's New Year Policy Speech) lay out an ambitious set of targets to be met by 2002, including: – Providing universal service access at speeds of 2 Mbps. – Connecting 10 400 schools to the Internet. – Teaching 900 000 civil servants, 10 million students and 600 000 military personnel to use computers.	• Korea has the potential to become a world leader in information infrastructure if it can introduce the necessary institutional and regulatory reforms to enhance competition in the telecommunications sector and implement the recommendations below. However, it needs to make a sharper distinction between the advantages of the production of ICT hardware and technologies and those of its effective use and application. There appears to be too much emphasis on the former and not enough on exploring the potential of the latter. • In general, more open competition and investment policies are required. Korea has the potential to become a regional logistics hub as a complement to the expansion of ICT services.

Regulation

• Korea and Japan are the only countries in the OECD that do not have an independent regulatory agency. This creates real and perceived problems of unfair treatment among industry operators.	• The Ministry of Information and Communications (MIC) has been trying to regulate markets and provide greater competition, but competition remains limited and there are concerns about preferential treatment.	• Korea needs to set up an independent regulatory agency outside the MIC. Priorities for improved regulation are: – Move to long-run average incremental costs to calculate interconnection costs; – Create a universal service fund; – Simplify the licensing regime and processes; – Price caps for dominant local loop provision; leased line and long-distance provision; and arrangements for the unbundling of services; and – Move to a more open system of allocating radio frequencies.

Competition and foreign investment

• There were very strong limits to foreign investment in the telecommunications sector.	• Korea has been liberalizing its telecommunications sector by moving toward complete removal of limitations on resale services. Up to 49% foreign ownership is now allowed.	• Eliminate restrictions on foreign investment in the telecommunications services and manufacturing sectors to enhance investment levels in order to gain from the globalization of ICTs and ICT services.

Local loop services

• Korea Telecom was the only telephone service provider until 1990 (a duopoly exists in long-distance calls).	• In 1997, Korea opened up the local loop market to competition. However, Korea Telecom's competitors must replicate all fixed-line investments, thus constraining real competition.	• Local loop services should be unbundled to stimulate competition and the introduction of new technologies such as ADSL.

Legislation and regulation for electronic commerce

• Prior to the crisis, there were no legislative provisions for electronic commerce, as this was still an incipient sector.	• Korea passed the Electronic Signature Law in July 1999. Outstanding issues include: restrictions on number of Internet financial services; and a requirement for Internet firms to have a physical presence in Korea. • The government has announced that the Door-to-door sales law will be repealed or amended.	• Korea needs to move to a new system of taxation and commercial regulation that provides for harmonization with evolving international rules, including prevention of tax avoidance/double taxation, ensuring privacy, and limiting transaction costs. • It also needs to reduce restrictions on Internet financial services.

Situation up to crisis	Ongoing reforms	Remaining issues
Electronic government • No major provision for electronic government was made before the crisis, as this is a relatively recent phenomenon.	• In the area of re-engineering and innovation in government through ICT, *Cyber Korea 21* lays out impressive targets for using ICTs to improve government performance (*e.g.* providing e-mail accounts to all public servants, digitizing public procurement, document circulation). • It has increased provision and delivery of social services on line; and helped the disadvantaged and disabled with online connections.	• Korea should pay more attention to innovation in government; to policy issues that cut across ministerial boundaries, consistent with the new role of government; and to improving capacity at the local level.

IV. Innovation system

Situation up to crisis	Ongoing reforms	Remaining issues

Profile of science, technology and innovation activities
• Korea's innovation system (KIS) is based on the catch-up model rather than on a KBE model that emphasizes innovation and building capabilities and linkages among key actors.
• The country reportedly spends 2.8% of GDP on R&D, but the efficiency of such investment is low as measured by scientific and technical articles produced per unit of GDP. Other issues include:
  – Low spending on effective basic research.
  – Little use of FDI and weak global linkages.
	• Funding for R&D has been increased from 3.7 to 4.1% of the budget, with emphasis on R&D in telecommunications, biotechnology and new materials. President Kim has announced that Korea will increase R&D's share in the national budget to 5% by 2003.
	• Government has begun an evaluation of the effectiveness of its R&D policy and has started to make changes.
	• The government has designated strategic fields for R&D support, especially in ICT areas, such as next-generation Internet, fiber-optic technology, digital broadcasting, wireless communications, software and computers.
	• The President established a National Science and Technology Council to co-ordinate the various ministries active in S&T policy.
		• While continuing to exploit the strengths of its catch-up model, Korea needs to move toward a new model with stronger university research capabilities, an enhanced basic R&D effort by the private sector, greater emphasis on diffusion efforts by the government, and more linkages among domestic actors and between them and international actors. Many of these steps are included in the new science and technology strategy for the year 2025 presented in May 2000.
		• The key challenge will lie in the implementation of the new model, particularly in terms of making the system more efficient, effective and interactive by strengthening linkages among participating institutions and actors, including those from abroad.
		• There is a need for a review of basic research capabilities in Korea in all scientific fields, according to international standards.

Universities • Universities do not carry out much scientific research, although research resources per researcher are above the EU average. The problem lies in the lack of incentives for research and lack of recognition of the value of research. Little emphasis is placed on basic research, due to low public funding and a dependence on private enterprise funding.	• The *Brain Korea 21* program aims at strengthening the research capability of the premier Korean universities.	• There needs to be increased government funding and greater emphasis on basic research at universities, as well as the creation of centers of excellence. • University regulations concerning teaching obligations, professors' evaluation and promotion, etc., that create obstacles to research work and contact with industry should be removed or changed.
Private sector R&D • Two-thirds of Korean R&D is undertaken by the private sector, and the bulk of this is carried out by large, vertically integrated *chaebol* that tend to internalize these activities and have few mutually reinforcing linkages with SMEs. • Most of the R&D undertaken by the *chaebol* serves to widen their focus rather to than deepen it, and little investment is made in basic research. Industrial R&D spending is concentrated in a few sectors.	• The *chaebol* have been downsizing their R&D efforts as a result of the crisis and are beginning to focus their R&D on core areas. • Some technology-based SMEs are investing in R&D, but they are still a very small minority.	• Korea needs to increase domestic and foreign competition through improvements in the overall economic incentive regime which will force the *chaebol* to reorient research toward greater specialization and to focus on their core competencies. • Financial and technical support to SMEs need to be strengthened in order to stimulate their investment in R&D and technological interactions with the *chaebol*, foreign firms and the GRIs. • SME technology, training and research policies should be better co-ordinated through the establishment of networks involving the regional and local authorities.

Situation up to crisis	Ongoing reforms	Remaining issues
Government policies • The Korean Government promoted R&D through various policies including tax and financial incentives, procurement, technical information, human resources, co-operative research, SME support, support for R&D commercialization, and public R&D labs. • There has been little objective evaluation. Despite some success stories, most national R&D programs and government research institutes have suffered from duplication with the private sector. • Overlap in R&D support and R&D programs among different ministries has been pervasive.	• The 1999 Law for the Establishment, Administration and Promotion of the GRIs has taken many of the GRIs out of the direct control of their ministries and made them accountable to the National Research Council Boards. Nevertheless, the processes for allocating funding on a more competitive basis and for making institutes more accountable for results are not sufficiently developed. • The government has begun to think about redefining the role of the GRIs.	• The government should complement market forces where R&D will yield highest social returns. • The public sector's R&D programs should place greater emphasis on diffusion and on strengthening systemic linkages. The government should: – Provide incentives for the GRIs to diffuse and commercialize their results. – Develop bridging institutions, such as a university-industry interface, specialized research firm spin-offs. • The government should redefine the role of the GRIs by: – Redirecting them toward more basic and long-term research through secure public funding. – Repositioning them with respect to universities and the private sector to reduce overlaps. • The government should move from sectoral promotion toward a cluster approach, as found in other OECD countries, and improve co-ordination between national and regional policies. • It should increase the mobility of human resources in S&T through the removal of regulatory obstacles that impair mobility of researchers between the public and private sectors; and encourage foreign research and management talent (as found in Singapore).
Weak global linkages • Korea has had both formal and informal barriers against inward FDI, and there have been few linkages with foreign public or private research. • Korea has not participated actively in international public research programs.	– The *chaebol* have been setting up research centers abroad as a way of directly tapping into foreign knowledge. – The *chaebol* have also developed some strategic alliances with foreign firms, but they need to develop more aggressive strategic alliances with technologically strong foreign firms. – Korea is beginning to develop co-operative and public research programs with other countries.	• It is necessary to re-examine policy measures in order to: – Encourage firms to tap effectively into the global knowledge base through FDI, international business alliances, and joint research programs. – Attract foreign scholars to research universities and institutes through scholarships and grants. • The government should use technology forecasting to obtain a better appreciation of trends in science and technology.

V. Strategy for knowledge-based development

Implementing the strategy • Prior to the crisis, a number of initiatives were taken by the government on issues affecting education, science and technology, etc. However, these were not integrated into a knowledge-based development strategy.	• A new vision for transforming Korea into a knowledge-based economy was announced by the President on 3 January 2000. • The task of mapping out the details of the strategy was given to the National Economic Advisory Council (NEAC), which was formed towards the end of 1999. • There is an encouraging trend toward building consensus in the policy design process through public hearings, consultations with wide-ranging think tanks and the private sector.	• In crafting the strategy and action plans, a number of issues should be taken into account, such as: – The role of government as facilitator and co-ordinator. – Reform should be systemic and involve design and implementation of measures which are consistent across different areas of policy making. – Build consensus and buy-in from key stakeholders. – Open up traditionalist attitudes for critical reflection among the public at large (*e.g.* education reforms, entrepreneurship, risk taking). – Address implementation concerns such as the respective responsibilities of agencies, institutions and private sector. – Address co-ordination issues with adequate resources and a clear focus. – Make provisions for monitoring and evaluation (*e.g.* by the NEAC). – Establish an institutional set-up and process to balance stability with the need for continuous adjustment of the strategy.

Chapter 1

The Knowledge Revolution:
A Challenge to Korea's Development Strategy

A. The current context

Korea has achieved one of the fastest rates of economic development in the world. Between 1966 and 1996, its per capita income grew by an average of 6.8% annually,[1] from under USD 100 in the 1950s to USD 10 550 by 1997. Life expectancy has risen to 72 years, just short of the OECD average of 76. Primary and secondary education were universal by the 1990s and, by 1996, Korea's tertiary enrolment rate was higher than for most OECD countries.

Towards the end of 1997, however, Korea experienced its worst economic crisis since the Korean War. As a result of the East Asian financial crisis, GDP contracted by almost 6% in 1998. Unemployment, which was less than 2.5% in the second quarter of 1997, rose to a peak of 8.6% in February 1999, and foreign exchange reserves fell to less than USD 5 billion in December 1997. Nonetheless, Korea made a remarkable recovery and the economy grew by 10.7% in 1999. The Korean Government expects the economy to grow by about 8% in 2000 and by around 6% in the following few years.

The recovery has its roots in many factors, including improving investor confidence in East Asia, and corrections to initial overshooting by the international capital markets. The government's macroeconomic adjustment policies and the accompanying structural reforms have been at the core of the recovery process: as a result of the crisis, the Korean Government began a series of significant reforms in the financial, industrial, labor, business environment and social sectors. Those that are most relevant for this report are summarized in the policy matrix attached to the Executive Summary, and are covered in the respective chapters.

Despite the very rapid recovery, there is concern about the prospects for sustained future growth. The 1997 economic shocks and 1998 crisis revealed fundamental structural weaknesses in the Korean economy. The country's input-driven growth strategy had already led to declining total factor productivity growth from the beginning of the decade.[2] In the financial sector, rapid credit growth based on weak credit evaluation and inadequate prudential regulation and supervision had made the banking system vulnerable to shocks. Conglomerates had become greatly overextended because of extensive diversification based on excessive leveraging. The concern over future growth also stems from an increasing awareness that the global marketplace is becoming more competitive and that radical changes are taking place in the production, exchange and use of goods and services, driven by what can be described as a knowledge revolution.

B. The knowledge and information revolution

Awareness that knowledge is a critical element of economic growth and increased welfare is not new. What is new is the speed of changes in the production and dissemination of knowledge made possible by the increase in scientific understanding and very rapid advances in information and communications technologies (ICTs). Advances in scientific understanding and in the codification of knowledge are facilitating the rapid development of new technologies. Scientists are now at the stage where they can begin to engineer materials at the molecular level and even life forms, and our economies are in the middle of a revolution in information, computing and communication technologies. The cost of voice transmission circuits has dropped by a factor of 10 000 over the last 20 years and computing power per dollar invested has risen by a factor of 10 000 over that same period. The ICT revolution, in turn, is increasing the speed and decreasing the cost of developing tools and instruments for basic research (prototypes, demonstrators, simulation techniques), as well as extending the power of electronic networks as research tools, permitting the generation of a larger set of new technologies.

At the same time, investment in ICTs has expanded.[3] Growth in connectivity has fueled the rise of the Internet. As more suppliers and customers go on line, the benefits of participating expand and the penalties for non-participation increase. The direct linkages between producers and customers imply a reduced role for intermediaries in many markets, including consumer goods, food, manufactured components and primary products. Entirely new markets are being created through networking (*e.g.* personalized information services, online securities trading). In addition, the impact of networking in the public sector has become enormous, with the potential for dramatic reductions in the cost of delivering services such as health and education, and with major effects on governance through the re-organisation of administrative and political institutions. The Internet is also likely to fundamentally alter social trends through its impacts on day-to-day activities, like working, shopping, communicating, and even leisure. The pervasiveness of these changes and their impact even at the macroeconomic level has given rise to the concept of the "new economy" (see Box 1.1).

The ICT revolution involves significant reallocation of jobs across industries, and changes in the skill and occupational composition of the workforce. In OECD countries, low-technology manufacturing jobs

---

Box 1.1. **The new economy**

In parallel with these developments, substantial changes in economic growth have fuelled the notion of a "new economy". First, there are observed changes in the pattern of economic growth. Attention has focused on the United States, which is currently experiencing its most prolonged boom ever. Although productivity performance continues to be exceptionally strong and unemployment has fallen way below what used to be viewed as compatible with price stability, there is little sign of inflationary pressure. In a number of other countries there is similarly surprisingly little evidence of inflationary pressures. At the same time, comparing across countries, it appears that the forces pushing toward convergence in growth rates have given way to increasing divergence. Broadly speaking, countries which are doing well seem to be gaining ground relative to the followers. There are signs of growing income differences within as well between countries (Wilson and Rodriguez, 1999; Analysys Inc., 2000).

Second, there appear to be important changes in the sources of growth. Traditional determinants of growth, such as macroeconomic policies and the functioning of labor markets, continue to affect the performance rates observed in the late 1990s. In parallel, however, there is a pervasive influence of "new factors". This applies most notably to investment in ICTs, which was long observed to show up "everywhere but in productivity figures". Today, the contribution of ICTs to output and productivity growth is visible, significant and rising in many OECD countries. In Canada, the United Kingdom and the United States, the growth contribution of ICT equipment amounts to about half of the entire growth contribution of fixed capital. In France, Germany and Japan, the contribution of ICTs to output growth has been somewhat smaller, but is still significant (Schreyer, 2000). A number of smaller OECD countries are witnessing similarly strong benefits. These developments appear to play an important role in changing price behavior, since improved access to information is accompanied by stronger competitive pressures at the micro level and reduced prices across a broad spectrum of industrial activities.

At the same time, investment in ICTs is not the whole story. On the contrary, it appears that the "new economy" has brought about a more complex interplay between different factors, with, at its core, the diffusion and use of knowledge. Among all OECD countries, those which have strong growth rates tend to demonstrate an upsurge in multi-factor productivity. While it is difficult to determine the underlying sources and mechanisms, the evidence points to the crucial interplay between investment in ICTs, changes in innovation processes, organisational changes and upgrading of human skills (OECD, 2000).

have declined, while the effects on jobs in high-technology manufacturing have been mixed. Overall employment in the OECD countries has been shifting away from manufacturing and into services. Within services, there has been strong growth in high-value-added service sectors such as finance, insurance, real estate, and business services, which tend to be the most skill- and information-intensive. As demand for low-skilled workers declines, the overall occupational structure of the workforce in OECD countries is moving toward higher skills in both industry and services. High-skilled, white-collar workers make up an increasing proportion of the workforce, and account for between 25 and 35% of total employment.

Significant investments have taken place in training; these are aimed at helping workers make use of the new technologies and perform new functions (OECD, 1998a, pp. 42-45). There are also important investments in other intangibles: expenditures on R&D in the OECD countries account for 2.2% of GDP on average, meaning that R&D investment makes up more than 10% of gross domestic investment. When other investments in intangibles, such as in patents and licensing, design, marketing, education, training and software are added, investment in intangibles may represent as much as gross domestic investment (see Chapter 6 for a discussion of these issues).

The rapid development and spread of knowledge facilitated by technical progress and the ICT revolution is creating a more competitive and interdependent world. The share of world trade (exports and imports) in world GDP increased from 28% in 1970 to 45% in 1997 (World Bank, 2000), indicating increased globalization and competition. Beyond trade, there is greater interdependency through an increase in foreign direct investments, international sourcing of production inputs and inter-firm alliances, including the internationalization of R&D and technology licensing activities. Increased availability of ICTs; deregulation of financial and product markets; and the liberalization of trade, investment, and capital movements are accelerating this process. Increased international competition in turn spurs firms to create new products and adopt more efficient production processes. The expansion of international trade and production provides firms with the resources to finance innovative efforts, especially in countries with small domestic markets.

The direct role of technology in this process is reflected in the changing patterns of international trade. Between 1976 and 1996, the share of high- and medium-technology products increased from 33 to 54% of total goods traded (Figure 1.1). On the other hand, the share of other primary commodities fell

Figure 1.1. **International trade by type of good, 1976 and 1996**

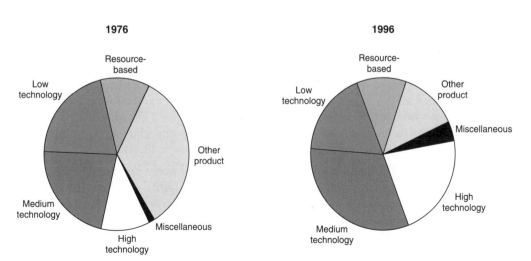

Note: The technology intensity of manufactured products is determined by the share of R&D to sales, where high-technology products are those with the highest shares.
Source: Adapted from World Bank (1999b), Figure 2.2, p. 28, based on World Bank Comtrade database.

from 34 to just 13%, while that of resource-based products remained constant. These trends have major implications for developing countries which are exporters of primary commodities. Not only the volume, but also the prices of their exports have been falling over the last five decades.

These figures, however, highlight only the most visible aspect of the relationship between technology and globalization. Technological change combines with increased economic interdependency to intensify and alter the nature of global competition across a widening spectrum of industries. In industries characterized as low- or medium-technology, the increase in technical knowledge and associated organisational changes increasingly provide an edge in productivity and enable product differentiation, significantly shaping competitiveness and value added. At the same time, globalization adds to pressures for adjustment and restructuring, and can thus adversely affect unskilled workers as well as firms in industries vulnerable to foreign competition.

While, in principle, the globalization of trade, finance and information flows may make it easier to narrow knowledge gaps across countries, the accelerating pace of change and the difficulties of many developing countries in getting started may, in effect, bring about the opposite result – a widening knowledge gap or "knowledge divide". If the knowledge gap were to widen, the world would be split even further, not only by disparities in capital and other resources, but also by disparities in knowledge. Increasingly, capital and other resources might flow to those countries with stronger knowledge bases, reinforcing inequality in a vicious circle (World Bank, 1999*b*).

There is also the danger of widening knowledge gaps within countries. The OECD economies are seriously concerned that the rapid advances in knowledge may be adversely affecting unskilled workers and increasing unemployment. In a number of countries, there is evidence to show that technology and technology-related organisational change is widening wage disparities between skilled and unskilled workers (OECD, 1998*b*). Such impacts are likely to be felt even more strongly in developing countries, where access to education and ICT infrastructure is far more differentiated and where formal safety nets are less prevalent. The rural areas and the poor are being left behind and thus run the risk of becoming marginalized in the knowledge-based economy (see Chapter 2 for a description of the situation relating to Korea).

C. Challenges to Korea's future development

Korea's rapid economic growth has been largely the result of high savings and investment rates. Studies of total factor productivity (TFP) growth indicate that although Korea's TFP has been respectable compared to that of other countries, it has been low compared to overall growth in outputs and inputs.[4] A recent study, for example, shows that of the 7.5% per capita growth between 1970 and 1995, 55% was due to growth in capital inputs, 20% to growth in labor inputs, and 25% – or nearly two percentage points – was due to TFP growth (Table 1.1).

Estimates of TFP depend very much on the methodology used for the calculations and most do not adjust for increasing returns to scale or to price mark-ups which are characteristic of markets with imperfect competition (such as those in Korea). When these adjustments are made, the contribution of TFP falls to less than 1%. Moreover, there is concern that per capita income growth will slow down as the contribution to growth due to increases in the working-age population and rising labor force participation, as well as from rapid increases in educational attainment, reaches its limits.[5]

A theme that will be developed throughout this report is that Korea needs to achieve higher productivity from its considerable investments in physical and human capital. This includes obtaining higher returns on the substantial resources invested in information infrastructure and in research and development (R&D). In the context of the rapid changes taking place as a result of the knowledge revolution and a very dynamic and interdependent global economy, achieving these higher returns will depend on redefining the role of the Korean Government and re-orienting its overall development strategy.

Korea's development model has largely relied on a government-led strategy based on the growth of large-scale industry. This model started in the early 1960s with the nationalization of key industries and banks. Control over the allocation of credit and a strong licensing and permit system allowed the

Table 1.1. **Total factor productivity growth estimates for Korea**

Contribution to GDP growth per capita per annum of 7.5% for 1970-95 (assuming constant returns to scale and competitive markets)	
Labor inputs per capita	20.5%
Capital inputs per capita	54.5%
Measured TFP	25.0%
GDP per capita	100%
Disaggregation of TFP growth of 3.6% per annum for 1975-95 (when adjustments are made for increasing returns and imperfect competition)	
Mark-ups	60.0%
Scale effects	17.9%
Technical change	17.3%
Other	4.8%

Source: World Bank (1999*a*), p. 47.

government to influence the industrial structure. The key focus in the 1960s was a strong export drive, which was successful in transforming Korea into an export-oriented economy. This was followed by the heavy and chemicals industry drive in the 1970s. During this period, there was an increasing concentration of economic activity in the *chaebol* as it was easier for the government to implement its policies through a smaller number of firms. These firms, however, grew larger and became more diversified, benefiting from privileged access to subsidized state finance and favorable regulatory and administrative interventions from the government.

In the 1980s, a series of policy measures were taken which aimed at strengthening market mechanisms. These included some trade liberalization, passing of the competition law, deregulation, and some privatization. Nevertheless, the role of the state and the reliance on the *chaebol* remained strong. Access to low-cost loans allowed the *chaebol* to continue their rapid expansion and to diversify into numerous industries – at the expense of profitability (Claessens *et al.*, 1998). This is corroborated by the declining rate of return on corporate assets in the 1990s.[6] In addition, despite declining profitability, the *chaebol* steadily increased their leverage ratios during the 1990s as they continued to access cheap government-controlled credit.

Korea's development paradigm is now at a crossroads. The development model predicated on input-driven growth and the rapid expansion of the *chaebol* had played its role and had begun to falter by the early 1990s. While the financial crisis in Korea was triggered by regional contagion, capital flight from the local currency and a weak financial system with poor prudential regulation and supervision, it was in large part also a corporate crisis, driven by excessive corporate debt and associated structural problems (OECD, 1999*a*).

Beyond its recovery, Korea is confronted with serious structural issues. According to the IMD *World Competitiveness Yearbook* ranking, Korea fell from 26th place in 1995 to 38th in 1999 as a result of weakening structural factors that culminated in the 1997/98 crisis.[7] For the year 2000, it moved back up to 28th place due to the many structural reforms that have been undertaken, although it continued to rank lower in finance, management, infrastructure, and internationalization. Meanwhile, there will be increasingly fierce international competition from advancing low-cost countries as well as in the advanced, knowledge-intensive areas.

The changing composition of manufactured exports from the East Asian economies highlights the challenges facing Korea. Over the period 1980 through 1997, Korea's share of world manufactured exports roughly doubled to about 3%. However, China increased its share of exports from about 0.5% of world manufactured exports in 1985 to nearly 4% in 1997, surpassing Korea, Singapore, Chinese Taipei and Hong Kong (China), the former developing country leaders (Figure 1.2).[8] Meanwhile, China has shown a dramatic fall in resource-based exports and a very significant increase in medium- and high-

Figure 1.2. **East Asian countries' share of world manufactured exports**

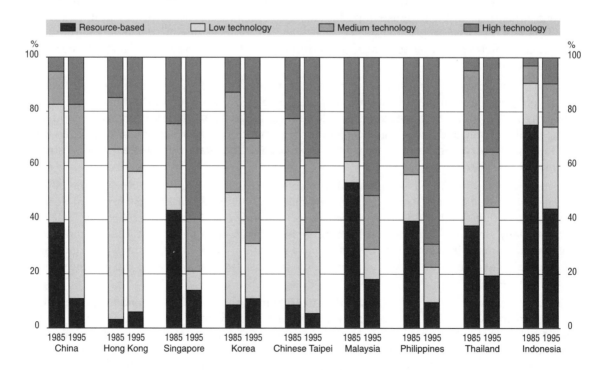

Source: UN Comtrade database.

Figure 1.3. **The changing composition of East Asian manufactured exports, 1985 and 1995**

Source: Lall *et al.* (1999).

technology industries (Figure 1.3).[9] China's very rapid upward shift is exerting significant pressure, not only on developing economy exporters of labor-intensive manufactures, but also on developing and advanced country exporters of medium- and high-technology products. All the other East Asian economies also show a significant expansion of medium- and high-technology exports, with the exception of Hong Kong, which has the largest share of low-technology exports. Singapore, Malaysia and the Philippines stand out due to their dramatic changes in export structures towards medium- and high-technology industries. These developments were largely enabled by export-oriented foreign investment, a strategy which, until recently, Korea had foregone because of its restrictions on direct foreign investment.

D. Korea's new development strategy

Caught between the rapid advance of the export-oriented developing countries in the region, and especially China, on the one hand, and the G7 countries, on the other, Korea is currently under strong pressure to shift its development strategy. Prior to the crisis, some analysts were already predicting that the growth of the Korean economy, and indeed that of the other East Asian economies, was not sustainable (Krugman, 1994). The Korean press highlighted the risks facing the nation in a series of reports centered on the theme of change and the need for productivity-based growth and a vision for the future based on knowledge.[10] The press also initiated discussions and dialogue with civil society and business groups on these issues. The government for its part, has intensified its efforts to develop an overall knowledge-based development strategy. This has included consultations with leading think tanks (the Korea Development Institute and others), as well as public hearings.

The task of mapping out the details of the strategy and the new vision, including that for the knowledge-based economy, was given to the National Economic Advisory Council (NEAC).[11] During its second meeting on 19 January 2000, the NEAC unveiled a proposal for a three-year master plan that will serve as the blueprint for Korea's transformation into a knowledge-based economy. The meeting, which was chaired by the President, presented three goals: i) make Korea one of the top-ten knowledge and information powers through a massive upgrading of the national information infrastructure; ii) improve the education system to meet OECD standards; and iii) enhance the Korean science and technology base to the level of the G7 nations.[12]

As Korea emerges from the crisis, now is the ideal time to take stock of where the country is, and what needs to be done to maintain a robust and sustainable recovery, remain competitive and increase the quality of life and welfare of its citizens in the 21st century. The remainder of this chapter develops the concept of the knowledge-based economy (KBE) and sets out a framework for analyzing the four main pillars that need to be addressed by Korea in its transition to an advanced knowledge-based economy. The four subsequent chapters summarize the key issues relating to each of these pillars. Chapter 6 sets out a program for the promotion of knowledge-based activities. Chapter 7 concludes with the importance of developing a consultative process for building national consensus to address the outstanding issues, and suggests elements for an integrated implementation strategy.

E. The concept of a knowledge-based economy

The OECD defines knowledge-based economies as "economies which are directly based on the production, distribution and use of knowledge and information". In fact, by that definition, all economies are knowledge-based. What is different, today, however, is that our rapidly growing economies are becoming increasingly dependent on the effective creation, acquisition, distribution and use of knowledge which is, to a large extent, enabled by the rapid advances of the science base and the ICT revolution. The effective use of knowledge and information is becoming the most important factor for international competitiveness, as well as for the creation of wealth and improved social well-being (Thurow, 1999).

The increasing spread of the "knowledge" sector in an economy can be measured in various ways. The pioneering attempt on this issue was made by Fritz Machlup in 1962. Machlup distinguished five broad categories of information activity: education, communications media, information machines, information services, and other information activities. He calculated that, in 1958, the knowledge sector accounted for about 29% of GDP in the United States, and about 32% of the labor force. Following the same methodology, Rubin and Taylor (1984) found that, by 1980, the share of the knowledge sector in GDP had increased to 32%. The OECD has also undertaken a number of studies of the knowledge economy, concluding that in the post-war period there has been a clear trend toward an economy where the share of the labor force engaged in the production, distribution and use of information is much greater than the share handling tangible goods (Foray and Lundvall, 1996, pp. 15-16).

Another way to trace the rise of the knowledge sector is to look at the share of various knowledge-based industries. The OECD defines knowledge-based industries and services as those which are relatively intensive in their inputs of technology and human capital. The narrowest category is high-technology manufacturing; that is, manufacturing where R&D accounts for a high percentage of sales. Its share of value added in the business sector in OECD countries averages 3 to 5%. A broader measure includes ICT manufacture and services. Using this measure, the share ranges from 5% in the United States to 10% in Japan, and has remained more or less constant, or fallen slightly, over the period 1985-94, reflecting in part the declining share of manufacturing in value added and a decrease in defense spending (OECD, 1998a). A still broader approach, which has been used most recently by the OECD, is to add communication services; finance, insurance, real estate and business services; and community, social, and personal services[13] (see Chapter 6).

Over the past several years, Korea has been developing a foundation for its knowledge-based or "new" economy by developing information and communications-related technologies, and by promoting computer literacy and the use of the Internet. Knowledge-based industries have played a leading role in Korea's economic growth. According to the Ministry of Finance and Economy (MOFE), between 1991 and 1999, the average annual real growth rate of Korea's knowledge-based industries was 13.7%, 9.6 percentage points higher than the 4.1% recorded by other industries.[14] In 1999, knowledge-based industries accounted for 45.6% of the nation's annual GDP growth rate. From 1991 to 1999, Korea's knowledge-based exports increased by an average of 26.4% annually, higher than the 11.3% recorded by other industries. These indicators reveal the increasing importance of knowledge-based industries. In recognition of this, the government is taking additional measures to optimize conditions for the knowledge-based economy. These measures are set out in the recently released "Three-year Information Technology Plan" and include: continuous expansion of investment in R&D and human capital; developing a high-speed communications network infrastructure; and encouraging foreign direct investment from technology-intensive companies (Ministry of Finance and Economy, 2000).

F. Definition and framework

Due to rapid technical progress, the development and application of ICTs and globalization, the acquisition and use of knowledge is becoming more central than ever to economic and social development. A knowledge-based economy is one that encourages its organisations and people to acquire, create, disseminate and use (codified and tacit) knowledge more effectively for greater economic and social development.[15] The KBE calls for:

- An economic and institutional regime that provides incentives for the efficient use of existing knowledge, the creation of new knowledge and entrepreneurship.[16]

- An educated and skilled population that can create and use knowledge.

- A dynamic information infrastructure that can facilitate the effective communication, dissemination and processing of information.

- A system of research centers, universities, think tanks, consultants, firms and other organisations that can tap into the growing stock of global knowledge, assimilate and adapt it to local needs, and create new knowledge.

These are the necessary conditions for increasing the efficiency, flexibility and resilience of the economy and its ability to restructure, respond to new challenges, permit the emergence of new firms and take advantage of new opportunities to ensure that the benefits of growth are broadly shared by society. The rationale for the focus on each of these factors is discussed below. The following four chapters will evaluate Korea's position in each and identify the key issues that need to be addressed in the transition to an advanced knowledge-based economy. Annex 1 benchmarks Korea against a group of 61 countries (28 developed and 33 developing countries), using a scorecard of basic indicators for each of the four areas. The highlights are summarized at the end of each section. Chapter 6 notes the interactions of these four pillars in terms of industrial structure and examines five structural policy areas that need to be addressed in promoting knowledge-based activities.

1. A *new economic incentive and institutional regime*

Whether or not the vast and rapidly growing stock of global knowledge is tapped and used efficiently by a country depends to a large extent on the economic incentive regime and institutional set-up in place. The incentive regime depends on the structural and market and non-market institutional arrangements within an economy. A competitive environment induces firms and individuals to seek out knowledge in order to produce goods or services more efficiently or to produce new goods and services. The general competitive environment and the pressures for more efficient firms to expand and for less efficient ones to shrink or disappear have a direct bearing on the diffusion of knowledge in an economy.

Taking advantage of the potential offered by the knowledge revolution requires a flexible society and economy, able to successfully cope with the need for constant change. This hinges on the existence of effective economic incentive regimes and institutions that promote and facilitate the constant redeployment of resources from less efficient to more efficient uses. This, in turn, means good macroeconomic, competition and regulatory policies and the existence of financial systems that can allocate resources to promising new opportunities (including venture capital) and re-deploy assets from failed enterprises to more productive uses. It also requires conditions that are conducive to entrepreneurship and risk-taking, the expansion of small enterprises, an adequate exchange of information between science and industry, and labor markets that are sufficiently flexible to ease the redeployment of labor. It implies the development of appropriate social safety nets to facilitate the constant relocation and retraining of people for new jobs, and to assist those people who are adversely affected by such restructuring.

The ways in which people access relevant knowledge and the incentives for them to gather, provide and use it are also affected by the institutional structure of a society. This involves relationships between legal rules and procedures, social conventions, organisations such as firms, and government and non-government organisations and markets. Equally important are the institutions that govern the rules and procedures in a society, which in turn determine how decisions are made and actions taken. A key feature is the quality of government, as its integrity and effectiveness determine the basic rules of a society. Another important element is the extent to which the legal system supports basic rules and property rights. The creation and dissemination of knowledge, for example, is strongly affected by the degree to which intellectual property is valued and its rights protected and enforced.

Finally, the ICT revolution will also have implications for governance systems which need to cope with the many demands that constant restructuring and redeployment place on the rules of the game, both at the international and the national level. Globalization and the trend toward a global innovation system is intensifying pressure for international harmonization in the treatment of intellectual property protection, competition policy and taxation. At the same time, the move towards greater democratization, transparency, accountability and decentralization in government, which are in part facilitated by the ICT revolution, will raise new challenges related to responding to local needs in an increasingly globalized environment. Compared to the most advanced countries, Korea still has some way to go on many aspects of the economic and institutional regime (see Annex 1). However, the Korean Government is using the pressures created by its recent financial crisis to undertake a major overhaul of its economic incentive and institutional regime and move toward a new knowledge-based development strategy. In Chapter 2, we analyze the progress made to date in the economic incentive and institutional regime, and highlight areas that still require attention. In Chapter 6, we analyze some of the key issues that need to be addressed to further promote knowledge-based activities.

2. *Skilled and creative human resources for the knowledge economy*

Skilled and adaptive people play a crucial role in taking advantage of the potential offered by the explosion of new knowledge and accelerating technical change. Ensuring that expenditures on education are allocated efficiently and that the entire population has the knowledge and skills necessary to participate in the knowledge-based economy, requires particular attention. Education is the basis for creating, acquiring, adapting, disseminating, sharing and using knowledge. Basic education increases peoples' capacity to learn and to use information. But this is just the beginning. Increasingly, it is also necessary to have technical secondary-level education, as well as higher education in engineering and scientific areas, to monitor technological trends, assess what is relevant for the firm or the economy and use the new technologies. The production of new knowledge and its adaptation to a particular economic setting is generally associated with higher-level teaching and research. In industrial economies, university research accounts for a large share of domestic R&D; in Korea, however, universities carry out very little research.

In the context of rapid development and dissemination of new knowledge, opportunities for life-long learning are essential. Creating a culture of continuous learning and openness to new ideas is critical for a knowledge-based economy. This should not be limited to learning on the job, but should be expanded to foster learning in multiple environments: at home, at school, and at work through structured continuing education courses, self-learning through the Internet, or computer-assisted instruction. In addition, in order to use the new technologies, and in particular the fast-evolving ICTs, workers and firms need new skills. Firms are investing heavily in training. Korean firms complain that they have to invest in retraining college graduates entering the workforce – an indication that, despite high levels of educational attainment, the educational system is failing to meet the needs of the knowledge-based economy.

In Chapter 3, we take a critical look at the educational and skills issues facing Korea from the perspective of the needs of the knowledge-based economy. Korea ranks favorably in educational attainment, even when compared to advanced countries (see Annex 1). However, the surprising conclusion is that, despite the very high levels of investment in education and skills, this is the area that requires the most fundamental reform and restructuring to put Korea in a position to take full advantage of the KBE.

3. A *dynamic information infrastructure*

The rapid advances in ICTs are dramatically affecting economic and social activities, as well as the acquisition, creation, dissemination and use of knowledge. These advances affect the way in which manufacturers, service providers and governments are organized, and how they perform their functions. Increased access to ICTs is affecting the ways in which people work, learn, play and communicate. As knowledge becomes an increasingly important element of competitiveness, use of ICTs is reducing transaction costs, time and space barriers, allowing the mass production of customized goods and services, and substituting for limited factors of production. With ICT use becoming all-pervasive and its impacts transformational, ICTs have become the backbone of the knowledge-based economy. To support Internet-based economic activities, countries need to ensure competitive pricing of Internet services and provide an appropriate legal infrastructure to deal with online transactions.

The National Information Infrastructure (NII) consists of telecommunications networks, strategic information systems, the policy and legal frameworks affecting their deployment, as well as the skilled human resources needed to develop and use the infrastructure. Developing a strong NII requires the mobilization of the many stakeholders involved in its deployment and use – government, business, individual users, telecommunications and information service providers, and so on. The formulation of an NII strategy involves identifying the opportunities and needs for information and communications infrastructure in the economy. It also involves assessing existing information systems in the economy to identify common constraints and difficulties, and formulating goals and visions for the NII. The latter are political roles around which to rally the resources of society. As such, the government has a very important role in the articulation of its vision, although this should be done in a consultative process with other stakeholders.

It is also essential to assess the knowledge and skills required to design, implement and use the new ICTs. This requires researchers and technicians across a spectrum of information technologies, a workforce that can use the new production technologies and a general population that can use electronic products, computers, software and information services effectively. Addressing these needs will require developing education and training policies, institutions and programs to prepare the appropriate human resources. While the government does not need to provide any of these services itself, it can play a key co-ordination role in identifying needs and weaknesses, ensuring that the necessary skills are developed and setting in motion the mechanisms to achieve these goals.

In Chapter 4, we analyze Korea's information infrastructure. In the scorecard indicators (see Annex 1), it became apparent that, based on data for 1997, Korea did not rank well compared to the advanced countries. However, the Korean Government has developed a comprehensive and ambitious plan with significant investments in this area, and much progress has been made to date. Nevertheless, there are still some major issues to be addressed in the regulatory and legal regime, which, if properly dealt with, could speed up progress in this critical area.

4. *An effective innovation system*

An innovation system consists of the network of institutions, rules and procedures that affect how a country acquires, creates, disseminates and uses knowledge (see Box 1.2). Currently, the bulk of technical knowledge is produced in the developed countries. Estimates show that roughly 88% of total R&D expenditures are undertaken by developed countries and these countries also account for roughly the same percentage of patenting or the production of scientific and technical papers. On a per capita basis, the disparity in the production of technical knowledge is greater than the disparity of income between developed and developing countries. Fortunately, developing countries do not have to reinvent the wheel: there are many ways in which they can tap into and use the knowledge created in developed countries.

Developing countries need to establish effective institutions in order to create, adopt and disseminate knowledge locally. Key components in the creation of knowledge include universities, public and private research centers, and policy think tanks. And, to the extent that they also produce new knowledge, non-government organisations, private firms and the government are also part of the innovation system. Some of the institutions central to the dissemination of knowledge include agricultural and industrial extension services, engineering consulting firms, economic and management consulting firms and, in the case of Korea, the GRIs. However, the mere existence of these organisations is not sufficient. What counts is the extent to which they are effective in creating, adapting and disseminating knowledge

---

Box 1.2. **The national innovation system**

There is no single accepted definition of a national innovation system; what is important is the web of interaction or the system as a whole. The concept of national innovation systems rests on the premise that understanding the linkages among the actors involved in innovation is key to improving a country's technological performance. Innovation and technical progress are the result of a complex set of relationships among actors producing, disseminating, acquiring and applying various kinds of knowledge. The innovative performance of a country depends to a large extent on how these actors relate to each other as elements of a collective system of knowledge acquisition, creation and use. These actors are primarily private enterprises, universities and public research institutes and the people within them. Linkages can take the form of joint research, personnel exchanges, cross-patenting, licensing of technology, purchase of equipment and a variety of other channels.

Source: Expanded definition based on that used in OECD (1997).

to the firms, government, other organisations and people who put it to use. Therefore, networking and interactions among the different organisations, firms and individuals are critically important. The intensity of these networks, as well as the incentives for acquiring, creating and sharing knowledge, are also influenced by the economic incentive regime in general. They are affected by specific policies relating, for example, to imports of foreign technology through technology licensing, direct foreign investment, foreign collaboration and policies on the protection of intellectual property.

In Chapter 5, we analyze these issues in the Korean context. According to official Korean data, the country actually spends more on R&D as a share of GDP than most other OECD countries (2.8% of GDP during 1985-95). However, the scorecard indicators (Annex 1) suggest that Korea does not appear to be getting full value for its money. Its broader innovation system must therefore be improved to take advantage, not only of the creation of new knowledge at home, but also for the acquisition of knowledge from abroad, and for the dissemination and effective use of knowledge regardless of where it is created.

Over the past few years, Korean industry has been fairly successful in its attempts to more effectively exploit knowledge in its economic activities. However, challenges remain. In Chapter 6, we outline the limitations to further improvements in Korea's industrial profile. To provide favorable conditions for the promotion of knowledge-based activities, the government will need to rectify a number of weaknesses in industry-related issues. Unless properly addressed, these weaknesses will dampen pressure for change and blunt the responsiveness of the private sector to new market openings. We also propose specific measures which could be used to promote knowledge-based activities, and highlight the trend among OECD countries toward providing an institutional set-up aimed at strengthening the capacity of industry and providing incentives for firms to respond to changing market conditions.

Given its high educational attainment, heavy investments in R&D, and very rapid progress in developing its information infrastructure, Korea is on its way to becoming a knowledge-based economy. Chapters 2-6 show that much work remains to be done to provide adequate conditions for this transition. Chapter 7 highlights some of the issues that need to be addressed in order to implement a successful development strategy.

Notes

1. In constant 1995 USD.

2. See World Bank (1999*a*), and Section C for details.

3. In a recent study of 60 major telecommunications markets worldwide (covering 90% of the telecommunications market), Pyramid Research estimated that in the 1991-94 period, communication service providers in those countries connected 118 million telephone lines, 40 million mobile subscribers and 6 million leased lines. Between 1995-98, these figures increased to 171 million new telephone lines, 238 million mobile subscribers and 8 million leased lines. Cumulative investments in ICT infrastructure in all the countries examined for the Pyramid study totalled USD 327 billion between 1991-94. Between 1995-98, such investments nearly doubled, to USD 572 billion. Pyramid Research also forecast a number of technological advances over the next ten years that will fuel rapid expansion in the virtual economy, with global consequences. As an example, the worldwide e-commerce market is forecast to expand from USD 26 billion in 1997 to USD 1 trillion by 2002. And *Cyber Korea* 21 predicts that the size of the e-commerce market will expand from KRW 55 billion to KRW 3.8 trillion between 1998 and 2002.

4. Total factor productivity (TFP) allows growth in value added to be broken down into the contribution of growth in capital and labor inputs and a residual that is attributed to technical change. The results of the decomposition depend on the type of production function assumed and on the degree of adjustments made to the quality of inputs (*e.g.* education of the labor force) or the quality and utilization rate of capital. For estimates of TFP for Korea, see Young (1993); McKinsey Global Institute (1998); World Bank (1999*a*).

5. One factor that still has strong potential is increasing labor force participation by women, an issue which will be developed in Chapter 2.

6. World Bank (1999*a*), pp. 46-47 and Figure 4.5.

7. The IMD methodology ranks a country based on structural factors and annual surveys of entrepreneurs. It covers eight factors: domestic economy, internationalization, government, finance, infrastructure, management, science and technology, and people.

8. It is striking that Hong Kong's share in world manufactured exports has actually fallen by half, to only about 0.6%, while all the other East Asian economies have increased their shares. This reflects its declining competitiveness in manufacturing as well as its shift toward a service economy. However, Singapore has also shifted towards a knowledge-intensive service economy, while at the same time increasing its share of manufactured exports.

9. Technology intensity is defined by the share of R&D to sales in each sector, as determined by an average R&D to sales ratio for the OECD countries. Although Korea has a stronger technological base than all other East Asian economies (with the exception of Japan), many other East Asian economies appear to show higher technological intensity in their exports because they rely on imported inputs with high technological components that are assembled locally and re-exported. Much of this is carried out by foreign firms using these countries as export platforms. There are also differences in product mix between Korea and the other countries.

10. The *Maeil Economic Daily*, for example, has sponsored three reports: Booz Allen Hamilton (1997); McKinsey (1998); and Monitor Company (1998).

11. The NEAC was legally created in November 1999 to focus mainly on future-oriented policy directions, in contrast to the more short-term policies that come under the purview of the Economic Policy Co-ordination Committee in the Prime Minister's Office. A further round of consultations took place within government agencies and research institutions to chart out the various action plans as part of the implementation strategy. This part of the process is expected to be completed in March 2000. Preliminary details of the plan were announced at the beginning of April 2000.

12. See Chapter 7 for the most recent developments in the institutional responsibilities surrounding implementation of the knowledge-based development strategy.

13. The financial sector is perhaps the most information-intensive sector in the economy. It aggregates information on the use of resources in order to allocate capital to the most efficient utilizations. It is also the most intensive user of ICTs and was the first to operate on a global dimension in real time, thanks to advances in ICTs.

14. The definition used by the MOFE appears to be based on the OECD's 1998 definition which includes high-technology manufacturing, communication services, and finance, insurance and business services, but excludes the two other categories – medium-high-technology manufacturing and community, social and personal services – which the OECD added in 1999. By this definition, according to the MOFE, knowledge-based industries in Korea increased from 14.7% of GDP in 1991 to 20.5% in 1999 (compared to 34% in the United States for the latter year).

15. The terms "new economy", "digital economy" and "knowledge economy" are currently used interchangeably, and usually focus on the creation and use of ICTs. In this report, we have chosen a broader definition of the knowledge-based economy that includes not only ICTs and industries dependent on high and medium R&D, but also covers the effective use of technical, policy and social knowledge for economic activities. However, this not as broad as other definitions, for example, that of the "knowledge society", which includes spiritual, social, intellectual and philosophical knowledge.

16. The economic institutional regime allows organisations, people and institutions to adjust to changing opportunities and demands in flexible and innovative ways. In a sense, it is the fundamental pillar of the knowledge-based economy, since in the absence of a strong economic incentive and institutional regime that deploys these resources to productive uses, it is possible to have a strong educational base and/or a highly developed R&D infrastructure (as is the case in Russia, for example), but to miss out on the full benefits of these achievements. Thus, it is the nature and quality of interactions among the four pillars of the framework that are important for the knowledge-based economy, as broadly defined in this report.

References

Analysys Inc. (2000),
The Networked Revolution and the Developing World, report prepared for infoDev, Washington, DC: World Bank, http://www.infodev.org/library/400.doc.

Booz Allen Hamilton (1997),
Revitalizing the Korean Economy Toward the 21st Century, Seoul : Maeil Business Newspaper.

Claessens, Stijn, S. Djankov, J. Fan, and L. Lang et al. (1998),
"Resolution of Corporate Distress: Evidence from East Asia's Financial Crisis", World Bank Policy Research Working Paper, No. 2017, Washington, DC: World Bank.

Foray, Dominique, and Bengt Åke Lundvall (1996),
"The Knowledge-based Economy: From the Economics of Knowledge to the Learning Economy ", in OECD, Employment and Growth in the Knowledge-based Economy, pp. 15-16, Paris: OECD.

Ministry of Finance and Economy (Korea) (2000),
Korea Economic Update, newsletter of the Korean Ministry of Finance and Economy, 20 April (Vol. 30, No. 5, p. 1).

Krugman, Paul (1994),
"The Myth of Asia's Miracle", Foreign Affairs, November.

Lall, Sanjaya, Manuel Albadejo, and Enrique Aldaz (1999),
"East Asian Exports: Competitiveness, Technological Structure and Strategies", paper prepared for ASEM (Asia Europe Meeting) Regional Economist's Workshop: From Recovery to Sustainable Development.

Machlup, F. (1962),
The Production and Distribution of Knowledge in the United States, Princeton: Princeton University Press.

Machlup, Fritz (1984),
Knowledge: Its Creation, Distribution and Economic Significance, Vols. 1, 2, 3. Princeton: Princeton University Press.

McKinsey Global Institute (1998),
Productivity-led Growth in Korea, Seoul, Washington: McKinsey Global Institute.

Monitor Company (1998),
Knowledge for Action.

OECD (1997),
"National Innovation Systems", free brochure, Paris: OECD.

OECD (1998a),
Science, Technology and Industry Outlook, Paris: OECD.

OECD (1998b),
Technology, Productivity and Job Creation – Best Policy Practices, Paris: OECD.

OECD (1999a),
Asia and the Global Crisis: The Industrial Dimension, Paris: OECD.

OECD (1999b),
Science, Technology and Industry Scoreboard 1999: Benchmarking Knowledge-Based Economies, Paris: OECD.

OECD (2000),
Is There a New Economy? The Changing Role of Innovation and Information Technology in Growth, Paris: OECD.

Rubin, M.R., and M. Taylor (1984),
The Knowledge Industry in the United States: 1960-1980, Princeton: Princeton University Press.

Shreyer, Paul (2000),
"The Contribution of Information of Communication Technology to Output Growth: A Study of the G7 Countries", STI Working Papers, 2000/2, Paris: OECD.

Thurow, Lester, C. (1999),
Building Wealth: New Rules for Individuals, Companies, Nations in a Knowledge Based Economy, New York: Harper Collins.

Wilson, E., and F. Rodriguez (1999),
  Are Poor Countries Losing the Internet Revolution?, report prepared for *info*Dev, University of Maryland, College Park.

World Bank (1999*a*),
  Korea, Establishing a New and Sustained Foundation for Sustained Growth, Washington, DC: World Bank.

World Bank (1999*b*),
  World Development Report 1998/99: *Knowledge for Development*, Washington, DC: World Bank.

World Bank (2000),
  World Development Indicators 2000, Washington, DC: World Bank.

Young, Alwyn (1993),
  "Lessons from East Asian NICs: A Contrarian View", NBER *Working Paper*, No. 4482, October, Cambridge, Mass.

Chapter 2

Updating the Economic Incentive and Institutional Regime

A. Introduction

The 1997 crisis clearly demonstrated the need to re-examine the economic incentive and institutional regime in Korea. The main challenge facing Korea is to move away from past interventionist policies to foster a flexible, adaptable, market-based networked economy and creative society.

The broad agenda presented in this chapter requires completing the structural reforms started in the wake of the crisis and putting these into practice, including redefining the role of government, creating conditions to support the entry of more new players in the economy, improving the soundness and efficiency of the financial system, increasing the flexibility of the labor market, preventing the potential "knowledge and digital divide", strengthening the basic institutional infrastructure and fortifying the rule of law.

B. Redefining the role of government

The Korean Government has played a key part in the country's rapid development, and will indeed play a critical role in the transition to the knowledge-based economy. However, this new role calls for the interventionist policies of the past to be abandoned. Meeting the requirements of the knowledge-based economy means making markets work more effectively. This will facilitate the constant redeployment of the resources that enter into the creation and implementation of new and better ways of doing things and the rejection of outmoded methods. However, it cannot be handled through heavy government intervention. The central elements of the changed role for government in this context are:

- Promoting competition and entrepreneurship, and deregulating the market to unleash the creative energy of individuals and the private sector, while at the same time providing a modern regulatory framework to support the efficient and equitable functioning of markets.

- Securing the rule of law and providing greater transparency, disclosure of information, and accountability for market players, as well as for government.

- Building and harnessing a modern legal and institutional infrastructure relevant for the knowledge economy, in such areas as intellectual property rights, valuation of intangible assets, cyber laws covering privacy, security and digital transactions.

- Addressing the access and equity issues that may result from greater reliance on stronger market forces, from continuous restructuring and from the risk of a growing digital divide.

- Providing public goods such as basic education and basic research.

- Addressing missing or underdeveloped markets and networks, including acting as a catalyzer for a high-speed Internet backbone,[1] promoting new technologies,[2] facilitating networking (*e.g.* between universities, researchers, and firms), while being careful to foster market-led mechanisms rather than impede them.

In addition, the government must modernize and restructure itself in order to capture efficiency gains by: improving the quality, efficiency and delivery of public services; strengthening government information flows; and building strategic information systems.[3] Since the Asian crisis, the Korean Government has made a very impressive start in this direction. Reforms in the corporate, financial and labor markets are being supported by a restructured public sector consisting of a "small, efficient and service-oriented government" which will be achieved through improved public sector management, including the tax system and administration; more transparency in expenditures and budget processes and human resource management; and greater outsourcing to the private sector.

Over the coming years, Korea will need to address a number of issues in this area. The program will entail completing government reforms to achieve a civil service that is knowledgeable and compatible with the information age through: an appropriate level of salaries; new modalities for policy making in the context of greater globalization; the creation of a knowledge-based government through extensive training, the building of government information networks and better methods of knowledge sharing;[4] and the definition of the role of government in knowledge-based activities. In addition, there is a need for privatization of state-owned enterprises, especially in the information infrastructure sector, in order to provide reliable, efficient and low-cost services through the effects of competition underpinned by the modern regulatory oversight found in other OECD countries. There is still a strong sentiment in the private sector that the government sector has not changed enough – it is primarily perceived as a regulatory institution rather than one that is supportive of individual creativity and fully engaged in encouraging the growth of the private sector. The government will have to become a facilitator of change, fostering broad participation consistent with the distributed power arrangements of the networked, knowledge-based economy.

While the government plays a critical role in creating the appropriate incentive regime and in strengthening the education and information base, entrepreneurship and innovation are central to the knowledge-based economy. These activities are carried out primarily by individuals, firms and non-government organisations. An essential element concerns spurring greater "technopreneurship", *i.e.* entrepreneurship combined with the use and development of ICTs in a broad span of sectors, including services. There are still fiscal and other barriers to entrepreneurship, however, and despite the rapid development of venture capital markets in recent years, challenges remain as regards the performance of new and small firms more generally and the degree of entrepreneurship among individuals (see further Chapter 6). A culture of entrepreneurship also involves a change in attitudes as regards failure. One reason for the high level of innovativeness and dynamism in the United States is that honest failure is socially acceptable. Many successful entrepreneurs failed several times in trying to bring innovative ideas to the market before finally succeeding. For Korea to achieve such a change in mind-set and foster a more entrepreneurial culture (World Bank, 1999) will require common action on the part of the education system, the media and society at large.

C. Opening up the economy and promoting competition

Opening up to trade: reducing tariff and non-tariff barriers. During the 1980s and 1990s, Korea embarked on a selective liberalization and deregulation of the economy. This resulted in a decline in the percentage of product markets in which the top-three firms together accounted for more than half of total sales: from 88% in 1981 to 76% in 1994. However, the degree of competition was relatively limited prior to the crisis, as evidenced by measures of the mark-up of prices over costs: relative to other countries, mark-ups were significantly higher in Korea in a large number of manufacturing sectors. Moreover, these mark-ups were found to be negatively correlated with TFP growth, suggesting that the lack of import competition had significant negative impacts on productive efficiency (World Bank, 1999).

Since the crisis, however, there has been further liberalization of the Korean trade regime. The simple effective tariff rate, which had declined from 6.8% in 1988 to about 4.5% in 1996, was brought down further in 1998 and, by 1999, 90% of the rates were under 10% and the effective tariff rate had fallen to 2.8%. Progress had also been made on non-tariff barriers and further deregulation is planned.

Opening up to foreign investment. After joining the OECD in 1996, Korea streamlined previously restrictive regulations and brought them up to internationally accepted levels. Since the crisis, it has undertaken major liberalization of foreign investment through the New Foreign Investment Promotion Act (1998, revised in 1999). As of May 2000, of a total 1 148 classified sectors, only four remained closed to FDI. The 14 partially closed sectors and three of the closed sectors are expected to be further liberalized in 2000. Market access has been expanded through the liberalization of sectoral restrictions and the facilitation of cross-border mergers and acquisitions. As a result of the liberalization and the requirement that over-leveraged domestic firms reduce debt equity ratios to 200% by the end of 1997, FDI more than doubled between 1997 and 1999 as domestic firms sold off assets (Figure 2.1). Most of the restricted sectors are in the service industries.[5] Some of the affected sectors that directly relate to knowledge diffusion are broadcasting, television and cable broadcasting, news agency activities, publishing of newspapers and periodicals, and telecommunications.

Korea now needs to focus on becoming more proactive in attracting the FDI that can contribute most to its present stage of development and on maximizing the positive spillovers from such investments in the economy, especially the knowledge-intensive industries and the service sector.[6] Much will depend on the extent to which the industrial structures, workforce competencies and institutional conditions, such as intellectual property rights, induce foreign firms to transfer skills, technologies and management practices, and allow domestic enterprises and individuals to capture spillover effects. The experience of countries such as Ireland and Singapore are particularly relevant in this area. Chapter 6 further develops some of these issues.

Deregulation. The first important step towards domestic regulation came with the passage of the Basic Act on Administrative Regulations in 1997. In 1998, the Regulatory Reform Committee was established under the Office of the Prime Minister. By July 1999, a total of 5 480 regulations had been eliminated, 5 695 had been streamlined and 7 707 regulations emanating from 934 subordinate laws had been revised. In addition, 1 466 out of 1 840 informal regulations had been abolished. Although significant progress has been made in liberalizing the trade and foreign investment regime, Korean markets are still encumbered with vague and complicated regulations, especially in the service sector. Excessive regulation is one of the reasons why the service sector remains relatively underdeveloped compared to countries such as Singapore, the United Kingdom, the United States and Japan (Figure 2.2).[7]

Figure 2.1. **Foreign direct investment in Korea, by sector**
USD million

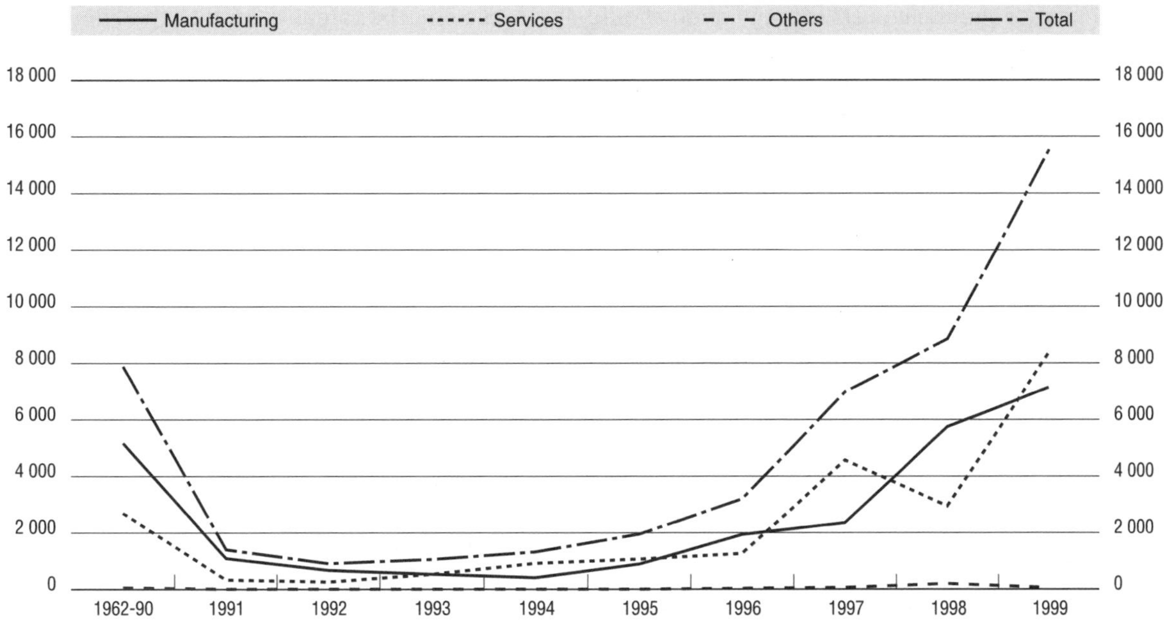

Figure 2.2. **Services as a percentage of GDP, 1995-98**

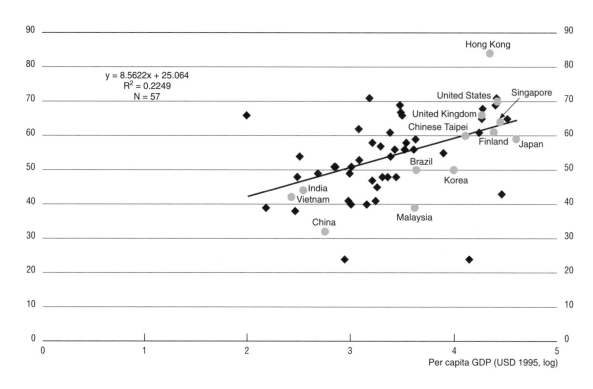

$y = 8.5622x + 25.064$
$R^2 = 0.2249$
$N = 57$

Per capita GDP (USD 1995, log)

Source: World Bank.

Maintaining the momentum of deregulation will be crucial in order to ensure the more competitive environment that would facilitate greater knowledge acquisition, diffusion and use. As will be developed in subsequent chapters, deregulation should also focus on other areas central to the knowledge economy – education and human resources, information infrastructure, and the science, technology and innovation system. While each Ministry is empowered to carry out its own deregulation mandate, overlapping areas have not yet been successfully resolved. Also, "informal guidance" should be underpinned by legal foundation and the regulatory oversight of infrastructure sectors should be brought up to the level of the OECD models. The strengthening of skills in the area of regulatory reform and modern regulatory oversight will be necessary as the country moves into the next phase of its transition to a knowledge-based economy.

Increasing domestic competition and entry. The recent trade and FDI liberalization and many of the current regulatory reforms should serve to improve the competitive structure of the economy. Korea's liberalization effort has in fact significantly increased the potential for foreign competition across the board. These actions will discipline industrial behavior, provide a more level playing field for new entrants and thus usher in a new industrial structure consistent with the development of the market economy. The challenge is to ensure that the reform effort is sustained and that continued progress is made in promoting a more competitive environment. The next step is to build capacity among regulatory institutions to shape a fair and efficient market system. The Korean Fair Trade Commission (KFTC) established in 1980 – which is in charge of enforcing the Monopoly Regulation and Fair Trade Act (FTA) – has been strengthened over the past few years.[8] In terms of overall competencies, the KFTC ranks well with other OECD and developed countries. However, to effectively fulfil its competition objectives, the KFTC needs strengthened capabilities in some areas, particularly with respect to investigative skills and enforcement capacities.[9]

Korea's industrial development and present structure has been much influenced by government support for the *chaebol*, and the impediments confronting foreign competitors as well as small and young companies in Korea. The government has implemented a range of measures to create a more favorable business environment, but there is still a strong element of command policies rather than measures which can put in place the mechanisms for spontaneous change. Also, much remains to be done with respect to the implementation and consolidation of existing policies. These issues are returned to in Chapter 6.

Bankruptcy and re-organisation. The Korean Government has reviewed measures to remove impediments to the effective implementation of bankruptcy and re-organisation laws. A New Bankruptcy Act has been passed and 99 companies from 21 *chaebol* have been placed under court supervision since 1997. Of these, 59 were placed under court receivership, which entailed a loss of ownership and management control, and 40 are in the process of re-organisation. Some asset transfers have also been completed, including the sale of Kia to Hyundai, or are under final negotiation. The Korean Government has been encouraging market-led workout programs. To improve the efficiency of restructuring under court supervision, in December 1999 the National Assembly passed amendments to the Korean insolvency laws, including the Corporate Re-organisation Act and Composition Act. Some important changes are: *i)* a shortening of the maximum period for courts to decide whether to accept a re-organisation petition, from five months to one month; *ii)* a reduction in the requirement for the approval of a re-organisation plan from 80 to 75% of the secured creditors; and *iii)* automatic application of the liquidation process in cases where a re-organisation plan fails. The amendment will be effective three months after its promulgation. The remaining issue here is the more effective implementation of the procedures and, over time, the move toward an increased reliance on market-based remedies through the legal system.

D. Increasing the soundness and efficiency of financial markets

Bank and non-banking institutions. Since the beginning of the crisis in November 1997, Korea has made significant progress in restructuring its financial system and reestablishing the solvency of banking institutions and has begun to lay the foundations for a sound and efficient financial system. Distressed banks are in the process of being restructured, viable banking institutions have been recapitalized, non-performing loans (NPLs) acquired from distressed institutions are being sold and a new regulatory and supervisory framework is being established. A major turnaround was made in terms of phasing out all policy loans, a cornerstone of the previous development strategy.[10] While these achievements are remarkable, the agenda for financial sector reform over the short term remains unfinished. The success of the financial sector restructuring program will depend upon the reduction of the still high leverage of domestic enterprises. A lack of significant progress in corporate sector restructuring will undermine any improvement in the health of financial institutions. Among the most important issues faced by the government in the short term are: resolution of NPLs; asset disposal by the Korea Asset Management Company (KAMCO); recapitalization of banking institutions; resolution and restructuring of Non Bank Financial Institutions (NBFIs), especially the merchant banks, finance companies, insurance and investment trust companies; and privatization of nationalized banks.

Despite significant improvements in the framework, additional changes are necessary to ensure that the financial sector promotes efficient mobilization and (re)allocation of funds to new and emerging knowledge-intensive sectors, and that markets and institutions exercise due diligence over corporations, including improved corporate governance practices. Moreover, improving accounting and disclosure standards, and minimizing moral hazard through the use of public funds for financial sector restructuring, remain a medium-term issue on the agenda.

While these actions would be necessary to improve the financial markets, the longer-term challenge is the creation of a sophisticated market-based financial system that will fundamentally alter the incentives and institutional framework influencing the behavior of banks and other financial institutions, enterprises and policy makers. Four sets of policies appear to hold the key to promoting a broader, more efficient and adaptable financial sector over time: continued strengthening of the incentive framework for the governance of financial institutions; promotion of securities markets (especially through

Table 2.1. **Total outstanding venture capital investments in Korea**

KRW billion

	December 1997	December 1998	December 1999
Total fund	2 501	2 606	4 691
Investment outstanding	1 833	1 511	2 188
% of GDP	0.40	0.34	0.44

Source: MOFE.

good corporate governance, accounting and legal frameworks);[11] further opening of the financial sector; and strengthening of legal codes and prudential regulation (see World Bank, 1999, for more details).

Venture capital. In Korea, the emphasis is placed on collateral-based debt rather than on equity. This practice presents a bias against small start-up firms in knowledge-based industries which can incur high costs for knowledge gathering, but which hold little physical collateral. Improving access to equity capital is important to the success of knowledge-based businesses.[12] Total venture capital investments outstanding amounted to KRW 2 188 billion at the end of 1999, or about 0.4% of GDP (Table 2.1).[13] In contrast, in the United States, outstanding venture capital investments amounted to about USD 95 billion, or over 1% of GDP.

To improve the venture capital market, Korea should strengthen regulatory oversight and information disclosure, ease restrictions on capital funds and open up to full participation of foreign investors. It is also important to further encourage a culture of entrepreneurship and "technopreneurship"[14] and to overcome the social stigma associated with business failure. Policies that promote the development of the public (stock) market may also provide an impetus for the (private) venture capital markets.

E. Enhancing labor flexibility and strengthening social safety nets

Enhancing labor market flexibility. The Korean labor market has been excessively rigid in the past, as measured by layoffs. Labor legislation, labor relations, and industrial and market structures have all contributed to labor market rigidities. Such rigidities hamper the ability of the country to respond rapidly to changing market needs and opportunities resulting from restructuring pressures and those brought about by the knowledge revolution.

A number of initiatives have been implemented since the crisis with the aim of creating a more flexible labor market. The government has revised the labor law to legalize layoffs; introduced a legal framework for manpower-leasing services; relaxed restrictions on private employment agencies; extended unemployment insurance to cover workers in all firms, including temporary and part-time workers; and introduced measures to improve the efficiency of active labor market policies. However, more work needs to be done in this area. The issues include overcoming labor's resistance to reform; revising employee benefits to make them fully portable; reorienting training schemes to meet the demands of the flexible economy of the 21st century (including the service sector) and multi-skilling of school-leavers and university graduates (see Chapter 3). A focus on these issues will certainly yield high payoffs in the knowledge-based economy through the creation of a labor force which is adaptable to new skill needs and lifelong learning. More recently, following the recovery, Korean labor unions have stepped up their opposition to further adjustments. Although a change in outlook has been achieved in other countries like Singapore, modifying the opposition of the labor unions to issues such as productivity, reskilling and retraining remains a challenge for Korea.

Strengthening social safety nets. Korea has also begun to strengthen its social safety nets – an important factor in enhancing labor market flexibility. As labor markets become more flexible, the risk of loss of employment and income for workers increases. The government has an important role in providing credible insurance against these risks, not only for reasons of equity but also to encourage appropriate risk-taking by individuals, thereby promoting labor market flexibility. Unemployment insurance, well-functioning pension systems and targeted poverty programs all form part of the overall insurance package. Since the crisis, the benefits structure and coverage of the unemployment insurance (UI) program have significantly improved. Coverage of UI now extends to all firms; the duration of benefits has been increased by two months for all categories of eligibility; and the minimum replacement rate has been raised from 50 to 70% of the minimum wage.

Korea has also embarked on a reform of its pension system. This has been motivated by several objectives: the social goal of securing adequate income support for the elderly; the fiscal need to ensure financial viability of the pension schemes; and the broader economic objectives of supporting capital market development and an efficient labor market.

Enhancing labor market flexibility is contingent on continued reform. Building on important steps that have been taken since the crisis, labor flexibility needs to be increased through better enforcement of existing legislation, further easing of legal restrictions, improvements in active labor market policies and the strengthening of social protection. In particular:

- Legislation on firing should be fully enforced to allow firms the flexibility to restructure in response to changing conditions.

- Restrictions that continue to apply to temporary workers should be removed to allow for a lower cost structure and the outsourcing of non-core activities. Besides the economic efficiency argument, there is an equity argument for removing these labor market rigidities which protect the interests of a minority of the workforce that is unionized at the expense of interests of the vast majority of the workers who are not unionized and who typically earn much less.

- Strengthening the social safety net will go some way to protecting workers from the higher frictional unemployment associated with more flexible markets. Although the payroll contribution rate of the UI was increased from 0.6 to 1% in 1998, it needs to be re-evaluated to assure its financial viability over the longer term. Similarly, the financial viability of the pension system needs to be guaranteed. Ensuring the portability of different pension schemes to allow workers to maintain their benefits when switching jobs is also crucial. One of the constraints to portability is the difference in the benefit formula applied to the NPS and the three occupational schemes. Thus, a comprehensive reform strategy needs to integrate the private and public pension schemes in a coherent framework.

F. Addressing the risks of the "digital divide"

The knowledge-based economy brings with it the risk that social inequality will increase due to rapid technological change, as a bias may be created toward specific production factors and because workers' preparedness for change differs.[15] Accordingly, Korea needs to pay more attention to the (re)training of workers, an area in which it has no significant institutional public program. The government needs to establish a comprehensive training/retraining plan to meet the requirements of the knowledge-based economy.[16] Sound industrial relations are also a prerequisite for productivity improvement.

Korea has rightly been hailed for its success in reducing poverty levels over the past three and more decades. Between 1965 and 1995, the absolute poverty rate declined from 41% to approximately 7%. With the economic and social crisis, which spread to the country in late 1997, poverty rates increased substantially, but the latest evidence (Kakwani and Prescott, 1999) is that these rates are once again on the decline.

As a result of the crisis, however, unemployment, poverty and inequality increased in Korea. The percentage of the population living in poverty increased from 8.5% in 1997 to 22.9% in the third quarter of 1998. Due to government income programs and the economic recovery, the rate came down sharply to 14.7% by the first quarter of 1999. Unemployment fell from a peak of 8.7% in February 1999 to 4.4% in November 1999. However, real wages, which fell by 12.5% from mid-1997 to the end of 1998, are still below their pre-crisis levels, despite substantial recovery in 1999. The Gini coefficient for urban incomes increased from .27 in 1997 to .30 in 1998. The National Statistical Office also reports that the ratio of incomes between the top and the bottom deciles of urban households widened to 8.5 in 1998 compared to 6.9 in 1997. There is also concern that many of the new jobs that have contributed to the falling unemployment rates are temporary and therefore less secure and less well paid. Women were significantly affected (see below) as they tended to be laid off before their male counterparts and, when rehired, their contracts tended to be short term.

Continued income disparities in Korea mean that the knowledge economy could create a "digital divide" such as that identified by the US Department of Commerce: a gap between those who have sufficient income and education to gain access to the potential benefits from IT, and those who do not.[17] During most of the period between 1965-95, the primary strategy for poverty reduction focused on the maintenance of macroeconomic stability in the context of steady economic growth. This approach has proven to be very successful in reducing overall poverty levels. However, even as overall levels of poverty continue to shrink, the hard-core poor remain poverty-stricken.[18] This group faces challenges that cannot be overcome through the somewhat indirect expedient of steady economic growth. The Korean Government recently recognized the limitations of the existing welfare system and is now developing a new "productive" welfare system which stresses the provision of basic education, medical and housing support. The rationale for the new system is to help the poor pull themselves out of the poverty trap by supporting their basic needs and implementing programs to develop their ability to work. New legislation, such as the Civil Basic Livelihood Secure Act, will become effective in October 2000.

Directions for future poverty reduction. A major question to be addressed is the potential impact of a knowledge-based economy on groups which are socially and economically vulnerable – the poor, particularly the rural poor; the disabled; and the uneducated – and the potential of ICT as a component of a poverty-reduction strategy. Such an analysis will make it possible to design better strategic approaches for assisting vulnerable groups to benefit from the knowledge-based economy and reduce their social and economic exclusion. The challenge facing Korea today is to understand the dimensions and nature of hard-core poverty, how this population is changing over time as a result of the economic and social transformations in Korea, and how best to design programs and interventions which can improve the lot of the poor in the context of these transformations.

Use of ICTs to assist targeted poverty reduction. The relationship between access to information and level of income is strong and becoming stronger at both the national and the international levels (US Department of Commerce, 1999; Thurow, 1999). The ICT revolution threatens to increase inequity, but it also provides tools to reduce poverty. Rural and poor urban communities can be integrated into economic life and their income levels raised through information services and electronic transactions. Appropriate regulatory incentives can be designed to encourage provision of rural telecommunications on a commercial basis. Satellite networks, wireless communications, public telephones and community information centers are effective arrangements. Public access "telecenters" can provide employment, training for school-children in basic computing skills, assistance in local government planning and co-ordination, support for maintenance of local ICT equipment, support for farmers (both price information and extension support), and outlets for the sale and production of electronic goods.[19] Information and communications technologies can also be used to monitor the poor and evaluate the extent to which such programs are affecting their lives.

Community information centers can extend the reach of information services to under-served rural and urban areas.[20] In the United States, a report on the "Digital Divide" recommended the extension of Internet access centers which provide ICT facilities for those who lack access at home and work. Other outreach measures being explored in the United States include: low-cost basic telephone service; cable connections for public institutions at little or no charge; regulatory mechanisms to enhance the availability of advanced telecommunications services to all educational and health-care institutions and libraries through mechanisms such as preferential rates for telecommunications services; and adoption of communications reform legislation which increases competition to make low-cost, high-quality services and equipment widely available, and to advance both rapid infrastructure modernization and expanded universal services for disadvantaged users, high cost or rural areas (see Annex 2). Such options could also be examined in the Korean context.[21]

Addressing employment biases against women. Despite considerable progress in the quantitative expansion of women's economic opportunities, including their increased access to education and training, improved health conditions, as reflected in significant declines in mortality rates and increased life expectancy, and advances in legal status, Korean women still lag behind men as regards opportunities for higher education and better employment. They continue to be discriminated against in recruitment and wage levels.[22] In addition to the moral argument against inequity of access by women to economic opportunities, there is also an economic argument. By not addressing such discrimination and by continuing to misallocate human capital resources that are vital for its competitiveness and sustained growth, Korea is missing out on the potential contribution of women to the workforce and hence to the nation's productivity. As noted in Chapter 1, the rate of growth is likely to be lower in Korea because the earlier contributions to productivity growth from increased educational attainment and increased labor force participation will essentially be exhausted unless there is an increase in educational attainment and labor force participation by Korean women. Korea ranks 24th out of 29 countries on the UN Gender-related Development Index (GDI) and 78th out of 102 countries in the Gender Empowerment Measure (GEM).

Overall, women are concentrated in services (60% of the total), agricultural and clerical occupations (47%), and constitute about one-third of professionals (32%).[23] The women's labor force participation rate in Korea was 47.4% at the end of 1999, down 2.1% from 1997. By comparison, men's labor force participation was 75.6%, down 0.4% over the previous year. In absolute numbers, more men than women lost their jobs due to the crisis. However, the percentage of the female labor force that was negatively affected by the crisis is higher than that of the male labor force. Women enter the labor market and become employed earlier in life than men but there is a tendency for women to withdraw from the labor force immediately after marriage, presumably for socio-cultural reasons (National Statistical Office, 1999, Tables 2.1 and 2.2). Support services for working women are limited: there are few childcare facilities and those that do exist are expensive. Many women, especially the educated, leave work after having children and choose to stay at home.

President Kim's Government is actively promoting greater participation of women in the workforce and in politics. The legal framework has been established in the form of the Equal Employment Opportunity Act of 1987, which has been amended three times, the last time in 1999 when the Anti-Discrimination Bill was introduced.[24] He also announced a new women's cabinet position in January 2000. However, despite the firm legal background and positive policies, women make up only 3% of national and local government, and discrimination against women in the workforce remains widespread.

Strategies for gender-related access to the benefits of the knowledge economy. Given that the gender equity problems that exist in Korea could potentially be exacerbated by the transition to a knowledge-based economy, affirmative measures in the workplace could include:[25]

- ICTs have the potential to open up alternative and more flexible working arrangements that would help women juggle their dual roles and reduce work stress on men. Introducing this flexibility into such a rigid work culture will be challenging, and should be done in such a way as to ensure that the benefits are felt by men as well as women.

- Opportunities for entrepreneurship through the Internet – used by women in other countries to enable them to work from their homes – should be explored. Linkages with international networks of women in business should be developed, and training provided for women entrepreneurs in Internet-based commerce. International linkages should also be developed through NGO networks such as the Women's Information Network for Asia and the Pacific.

- Provision of improved, less costly childcare facilities, including those in private sector enterprises.

- Consensus-building between the public and private sectors, labor unions and women's organisations on enhancing the contribution of Korean women to the economy and society.

- Developing consensus-based measures to eliminate discrimination in personnel management systems including promotion and wages, using the example of efforts made in this area by leading private and public sector industries in developed economies.

- Giving public recognition to firms and government agencies employing substantial numbers of women in professional and managerial positions.

- Supporting policy initiatives on new ways of doing business, including telecommuting and other flexible arrangements such as working from home, part-time work and job-sharing, global sourcing arrangements, new training and education capabilities.[26] In a survey of women-owned enterprises conducted by the Small and Medium Business Administration between October and November 1999, it was found that over 70% of women-owned enterprises had personal computers (PCs) and on average each enterprise had 4.4. computers; 11.1% of women-owned enterprises managed their own homepage. However, the same survey showed that 64% of women business owners used PCs only for simple office work, 21.4% of them used the computer for Internet search, while 44.4% of women entrepreneurs are computer illiterate (SMBA, 1999).

- Promoting the utilization of such innovations, particularly among women, the poor and other vulnerable groups, through outreach efforts involving a wide range of stakeholder organisations and networks, particularly targeting rural areas.

G. Updating the institutional regime

The increasing importance of knowledge and the greater ease of acquiring, processing and disseminating information are creating pressures to update many areas of the legal and regulatory structures in order to enhance the flexibility of markets and provide incentives for the creation, diffusion and use of knowledge. An area in need of immediate attention is intellectual property. However, more generally, reforms are also necessary to the institutional regime pertaining to information disclosure; transparency; accountability; rule of law, as well as the structure and functioning of government, including issues of governance and the reduction of corruption.

Strengthening intellectual property rights laws and their enforcement. The protection of intellectual property rights (IPRs) is becoming increasingly important in knowledge-based economies.[27] This is being driven by the mounting costs of R&D for new products or processes, the shortening of the product life cycle, the rapid growth in international trade in high-technology products and the internationalization of the research process.[28] This trend presents a major challenge for Korea as it moves from the knowledge acquisition and adaptation phase of the past, to becoming a generator and user of new knowledge and technologies. In the 1999 World Competitiveness Index produced annually by IMD, Korea ranked 41st out of 47 countries in terms of patent and copyright protection.

Quite apart from the obligations of global treaties that deal with IPRs (the TRIPs agreement), the incentive to develop knowledge is weakened if that knowledge is not protected. Patents and copyrights are a compromise between the need to provide an incentive for the creation of new knowledge and the social goal of disseminating knowledge at little or no cost. Giving property rights for a specified period of time provides the incentive to create the new knowledge. Requiring the publication of the knowledge in question partially fulfills the objective of announcing to the public that the knowledge has been created.

IPRs are important because: i) they provide incentives for innovation that cover the costs of creation; ii) they are an essential vehicle for transferring technologies and investment in high-technology industries, if the correct price is paid for sharing technologies; and iii) they help to attract funding for innovation and knowledge creation – sharing the risks and costs of R&D and encouraging collaborative research between domestic firms, universities and research institutions, and between domestic and foreign research consortia.[29] All of these factors are very relevant for Korea, given its strong R&D capability, its advanced industrial structure and its strong desire to become a knowledge-based economy. Intellectual property rights create bundles of rights that lend themselves to efficient exploitation through the market – production responsibilities can be divided among large and small firms and businesses – which is essential for the competitive development of Korea's high-technology small and medium businesses.

It used to be a generally held belief among Koreans that information and knowledge are public goods, thereby constraining the incentives for knowledge creation and provision. This thinking has been changing recently with strong support given to IPR issues. Korea is classified with "Priority watch list" status under the special 301 provision of the US Office of the Special Trade Representative, which does an annual classification of intellectual property rights regimes as they affect US trade and investment. Advocacy and education in IPR issues has been advancing. The Korean Government recently took the stance of valuing intellectual properties – technologies, knowledge, intellectual assets, and so on – and using them as materials for evaluating companies (and their viability),[30] and for new ways of calculating GNP. The government is also cognizant of the fact that, with globalization and the spread of ICTs, technical disputes in high-value technologies have emerged between domestic firms and between domestic and foreign firms.

Recent amendments in legislation should improve the overall protection and enforcement of IPRs in Korea. In particular, the Patent Court and the amendments to the Trademark Act (TMA) have been established, and both took effect in March 1998. These have already yielded positive changes.[31] The media is also playing a positive role in publicizing the enforcement efforts and in reinforcing governmental efforts. The bankruptcy of Hangul and Computer Co. (HCC) gave a valuable lesson that if intellectual property is not respected and protected, knowledge-creating activities will not be sustained. HCC had been the leader in Korea's word-processing software market until its bankruptcy in 1998. Even Microsoft, with its aggressive marketing campaign, could not outpace HCC. One of the reasons for HCC's bankruptcy was the frequency of pirated copies. Following the news of HCC's bankruptcy, the Korean public initiated voluntary campaigns to buy original products, and these have helped to resuscitate the Hangul software.

Despite the efforts that have been undertaken so far in this area, a number of outstanding issues remain:

- A sizeable portion of the public does not understand the criminal nature of actions which amount to infringement. Advocacy of IPRs to demonstrate their value to firms and their role in the stimulation of intellectual creation therefore needs to be strengthened. This has to be done jointly through co-ordinated efforts among government agencies, and through collaboration between government and private research institutes and the media.

- The Korean Government should increase public information services to enable better use of and access to new knowledge, whether patented or not. It should increase its efforts to produce more effective documentation and to compile knowledge in such areas through databases. The quality of institutions providing information services, such as the Korea Institute of Industry and Technology Information (KINITI), should be upgraded.

- There is a need to further extend the IPR laws and the administration of the new IPR laws – copyright protection – to pre-existing works and sound recordings (as part of the TRIPS Agreement); patent protection for pharmaceuticals and the protection of data for pharmaceuticals; protection of trademarks; software protection (there is a need to amend the Computer Program Protection Act); improved implementation of the Patent Act and Utility Model Act to streamline the examination and appellate process; and protection of trade secrets that are divulged by foreign businesses in business plans or on products as part of the requirements for entry to Korean markets.

- Current administrative structures for protecting IPRs need to be streamlined. The Korean Intellectual Property Office (formerly the Korean Patent Office) is in charge of patents, utilities and trademarks; the Ministry of Culture is responsible for copyright; the Ministry of Information and Communication (MOIC) is in charge of computer programs; and the Ministry of Agriculture is responsible for plant seeds.

- The courts, prosecutors, police and other government agencies have increased enforcement of IPR laws and are maintaining or increasing funding and resources to combat IPR infringement. Protection of IPRs can be improved through progressively oriented judicial activism in the interpretation of the laws, through enforcement of the laws, and through the proactive involvement of the prosecutor's office and other government agencies.

Rule of law, fairness, anti-corruption campaign, equal opportunity. Following the February 1998 elections, the Korean Government embarked on an ambitious program to reduce corruption in the public sector. On 17 August 1999, a comprehensive package of anti-corruption measures was announced. The package includes specific targets or measures such as increasing the salaries of public servants so that they match those of private industries within five years, compensating whistleblowers with cash rewards amounting to 5 to 15% of the government's recovered revenues, launching inspections of large local governments such as those in metropolitan cities and provinces upon the request of 1 000 or more citizens, and making it obligatory provide information on the content and use of political funds. As an indicator of a just and fair society, the government aims to become "transparent" by 2003, which implies a considerable improvement from the country's current 43rd position to 20th position in the ranking of Transparency International.[32] Progress is also indicated by the growth of many new non-government organisations (NGOs) which are pushing for greater transparency and accountability for government and society at large. The issue here is the timely implementation of the measures, including the announcement of the scorecard of success. In addition, upholding the rule of law to protect the poor, women and the disabled will be important.

Norms, standards, consumer protection. A knowledge-based economy is predicated on norms and standards that facilitate production and transactions while protecting consumers. This is an area where Korea needs to consolidate its strengths. As in other East Asian economies, the Korean regime was tilted in favor of production compared to transactions and consumption. The typical sequence in the industrial sector was to buy standard hardware technologies from abroad, use and adapt these technologies, and then follow-up with standardized production and exports. The scope and domestic development of (product) standards, certification, codification and accreditation were limited. Building such capabilities in the public and private sectors and moving towards international benchmarks in these areas will be crucial for Korea's next phase of quality- and innovation-based competitiveness and will also assist in the greater diffusion and commercialization of technologies. The role, functions and actions of the Consumer Protection Bureau need to be better understood, including its legal and institutional framework. Although the Consumer Protection Act was passed in 1980, it was only after it was revised in 1987 with the establishment of the Consumer Protection Board that attention was paid to the issue. While the Board has investigative powers, its capacity needs to be strengthened for advocacy, testing for consumer safety and investigative functions. Product information disclosures have been made mandatory.

In 1999, more than 300 cases of unfair advertising were filed with the Board and more cases are surfacing in the service sector relating to finance, e-commerce and cellular phones. The generic issue here is one of asymmetric information between consumers and sellers. A related issue is co-ordination of the functions that are now shared between the KFTC, MOFE and MOCIE. The formation of the National Consumer Policy Review Committee, headed by MOFE and composed of ministers, representatives of consumer groups and professors, is a step in the right direction.

H. Conclusion

To achieve a significant shift toward the knowledge economy, it is necessary for Korea to undertake a broad-based and interlinked reform package that covers the four pillars developed in Chapter 1. However, it is suggested that the highest priority be accorded to updating the incentive and institutional regime: this is critical to enable Korea to position itself on a new and sustainable growth path. These reforms are central to the overall functioning and flexibility required by the knowledge-based economy and the other three components of the framework. For example, pressing on with the reform of the incentive regime will induce the *chaebol* to become leaner, transform themselves into knowledge-based entities, make them more competitive in terms of innovation and encourage them to increase the productivity of their R&D efforts. This will promote Korean "technopreneurship" and at the same time create a market for the valuation of intangible assets. The incentive and institutional regime changes discussed in this chapter would also open up the education, training and information infrastructure markets. Chapter 3 discusses the education and human resource needs for Korea's transition to the knowledge-based economy.

Notes

1. In large infrastructures, such as the high-speed Internet backbone, profits depend on economies of scale and scope and on complex arrangements influenced by laws and regulations, and benefits may consist of public goods that are not captured as profits.

2. Including, for example, policies to encourage the use of computers and Internet among the population.

3. This includes sector-wide information systems for education, health, public sector management, transport, electronic payments, university and science networks, trade facilitation, property and business registries, disaster prevention and national statistics.

4. Some of detailed measures contemplated by the government include: *i)* strengthening co-operation between government ministries by raising the level of the Ministers of Finance and Economy, and of Education to that of deputy Prime Minister in charge of co-ordinating other ministries, a development committee for human resources and a ministers' meeting on welfare policies, voluntary reengineering of each ministry; *ii)* accomplishing e-government in the short term, through a knowledge management system and the introduction of e-administration (e-mail reports, electronic approval systems and electronic data interchange and training; civil affairs office to provide 24 hour service over the Internet); *iii)* improving the budget management institutions (by increasing the efficiency of public spending, including that of local governments); *iv)* proceeding with deregulation; and *v)* reforming state-owned enterprises and privatization (Ministry of Finance and Economy, 2000).

5. The 14 partially restricted sectors are: cereal grains production, newspaper publishing, publishing of periodicals, coastal water passenger transport, coastal water freight transport, scheduled air transport, non-scheduled air transport, wire telecommunications, wireless telecommunications, domestic banking, trust companies, cable broadcasting, and electric power generation (Ministry of Commerce, Industry and Energy, 1999).

6. In discussions with KOTRA, ambiguous tax laws and cumbersome regulations on customs and import procedures were identified as the most serious impediment to foreign investment.

7. In the retail trade, retail banking, telecommunications, and construction sectors, for example, product market regulations appear to have limited the degree of competition (McKinsey Global Institute, 1998). In the telecommunications sector, although long-distance and the international services were deregulated, government interventions on pricing decisions have limited the ability of new entrants to compete against the incumbent on price, and a lack of equal access to subscribers has further impeded their efforts to compete. The deregulation of the services sector is particularly important, given the large productivity gains that can be reaped through deregulation (in the US communications sector, for example, deregulation has brought productivity gains that have exceeded the economy-wide average by two-thirds over the last 20 years) and the important synergies between the manufacturing and services sectors that can be exploited. Many such anti-competitive regulations are now being eliminated. The government has also announced a privatization program, the first phase of which includes Korea Electric Power (KEPCO) and Korea Telecom (KT). The intention is to introduce effective competition, and the Ministry of Budget and Planning recognizes that regulatory oversight will need to be developed before any significant privatization transaction takes place.

8. So far, the KFTC has focused largely on unfair trade practices, which include predatory pricing and cross-subsidization of subsidiaries, false advertising and refusal to sell. It has done less in the areas of abuse of dominant position and resale price maintenance, areas which require more analytical skills. Also, monitoring of mergers and acquisitions (M&As) has been generally limited to responding to pre-notifications. Capacity needs to be built to assess more complex M&A cases where enhanced efficiencies of M&A outweigh the effects of competition restriction.

9. For more detail, see World Bank (1999) and OECD (2000).

10. Policy loans were given under the directives of the Ministry of Finance to selected sectors, especially to the heavy and chemical industry sectors.

11. Financial – and in particular capital market – development appears to be positively affected by good corporate governance. Korea has made great strides in strengthening corporate governance since the crisis. The issues involved fall in the broad categories of: *i)* shareholders' rights, particularly those of minority shareholders;

ii) institutional and strategic investors' rights; *iii*) creditors' rights; *iv*) accountability and functioning of corporate Boards of Directors; *v*) improving the legal system, including better insolvency procedures; and *vi*) improving the transparency of financial information and ensuring accountability (see Chapter 6 of this report and World Bank, 1999, for a more detailed discussion). As these reforms take hold and are deepened, they should provide further impetus to the development of Korea's stock markets.

12. Venture capital has enabled the growth of new knowledge-based industries in sectors ranging from computer technology to pharmaceuticals to biotechnology. The explosion in the Internet and telecommunications in the United States was facilitated by the presence of robust venture capital. Venture capital provides the financial foundation for innovation that is driving knowledge-based economies.

13. Even in larger industrial countries the importance of venture capital varies significantly. In the United States, Canada and the United Kingdom, where venture capital markets are the largest, new venture capital investments averaged 0.15-0.185% of GDP. In many other countries, such as Germany, France or Italy, however, annual venture capital investments have been of the order of 0.04-0.06% of GDP.

14. Some countries, such as the United Kingdom, have explicitly made an effort to teach entrepreneurship in schools. See UK Department of Trade and Industry (1998). Singapore has since the early 1990s engaged in the teaching of "technopreneurship" at the National University of Singapore. See *http://www.ec.gov.sg* for initiatives in this area.

15. This is also the case for Korea. During the 1980s, technological change widened the educational wage differential across industries (Choi, 1993).

16. For example, in collaboration with universities, government could introduce study courses for workers. Austria utilizes its *Fachhochschulen* system to this end. Finland also uses public/private partnership programs to enhance professional education.

17. Other factors that can lead to such a gap are geographical location and language barriers and outmoded mindsets.

18. The share of the Korean population that is below the international poverty line of USD 1 per day is less than 2%. See World Bank (2000).

19. There are also some inherent poverty and income inequality issues in relation to access: computer technology is expensive, with the result that unless specific action is taken to redress the imbalance, the rich will be in a better position than the poor to take up the opportunities provided by ICT. To increase the alternatives available to poor and vulnerable groups, parallel initiatives will need to be undertaken in the areas of small and medium enterprise development, access to credit, service and tourist industries, and cottage industry development and marketing.

20. In Amsterdam in 1994, a government-sponsored project created the Digital City (*De Digitale Stad*) as a virtual market and town center to facilitate community participation in government. In the first few weeks, 100 000 people joined. In North America, Freenets with similar purposes have been created in about 50 cities. With government support, the effort is being extended to rural areas of the United States. The Nordic telecottages provide isolated village communities with access to information resources for distance working, distance education and other capabilities. The Community Learning Utility project provides distance learning in Europe, and similar initiatives have started for some developing countries.

21. In March 2000, the Korean Government launched a nationwide campaign to teach Internet to one million housewives over a period of 18 months. In Seoul and nearby cities, nearly 70% of private computer institutes joined this program and are fully booked through July 2000. The program provides 20 hours of Internet courses per month for just KRW 30 000, far below the going market rate of KRW 100 000.

22. While the average life expectancy for men in Korea increased to 67.2 years in 1990-95, nonetheless Korean men have a shorter life expectancy than men in other OECD countries; Japanese men live 8 years longer on average. Meanwhile Korean (and Japanese) women have life expectancies (currently 74.8 years), that are among the highest in the world, rivaling and sometimes exceeding those in OECD countries.

23. National Statistical Office (1999), Table 9. The status and occupation of a worker is also differentiated by gender. Fewer women than men are self-employed (20% compared to 34% of men), and many more are unpaid family workers (22% of women compared to 2% of men). The distribution of employees is similarly gender-differentiated: a much higher proportion of women (43%) are in temporary employment (the share for men is 21%).

24. Women's wages are lower than men's despite the 1987 Equal Opportunity Act which guarantees equality between men and women in employment, and its 1989 revisions which include provisions of equal pay for equal work regardless of sex. In 1992, the average monthly wage of female workers was only about 56% of those of males, although this figure represents a 12 percentage point increase since 1972.

25. Initiatives that can be taken at the school and university level are addressed in Chapter 4.

26. In much of the industrialized world and specifically in the United States, it is largely the private sector which creates the information, develops applications and services, constructs facilities, and trains others to tap its

potential. All these areas offer flexible employment opportunities for Korean women. The need for adequate technical support can provide considerable job openings for women trainees, including flexible working hours which could accommodate their dual roles. The involvement of women's networks in the non-governmental sector, in academia and among students should be actively sought to promote take-up of such training.

27. Intellectual property consists of creative works, inventions, trade secrets and business goodwill. The scope and nature of intellectual property rights vary depending upon the rights concerned. Some of the rights protect ideas, some the expression of the idea, some rights give the limited form of monopoly right, while others merely give a right to prevent copying, and so on. In general there are seven types of rights – patent, copyright, registered trademarks, registered designs, plant variety rights, confidential information and circuit layouts.

28. At the conclusion of the 1994 Uruguay Round of multilateral trade negotiations that led to the creation of the World Trade Organization (WTO), a new agreement on trade-related aspects of intellectual property rights (TRIPs) strengthened IPRs in WTO member countries and gave developing countries a transition period to make the adjustment.

29. A study of Brazil and Mexico showed that in both countries, prior to the strengthening of IPRs, researchers tended to leave their countries as soon as research results became promising and take out patents in developed countries offering adequate IPR protection. Venture capital start-up was slow to develop as small start-up businesses feared that the venture capital firms might appropriate their knowledge; and knowledge interchange among researchers and technicians was constrained as researchers had no way of protecting their findings (Park, 1998).

30. In 1977, the Korea Technology Guarantee Fund (KOTEC) created the Technology Appraisal Centers in Seoul, Pusan, Taejon and Suwon to evaluate technologies owned by venture start-up companies. Made up of teams of experts in engineering, business and accounting, the Centers evaluate technologies owned by companies in terms of monetary value and business prospects. These assessments are used to provide credit guarantees by KOTEC or as collateral for loans from financial institutions. Patent rights, design rights, IPRs and commercialized technologies are appraised.

31. The amendments to the TMA were intended to streamline registration and recording procedures and to meet the necessary international standards for Korea's accession to the Trademark Law Treaty adopted by the WIPO. KIPO has mounted its advocacy function through education and training.

32. In recent years there has been a surge of interest in the consequences of governance for development. In a recent paper, Kaufmann *et al.* (1999) construct six aggregate indicators corresponding to six fundamental governance concepts. The authors define governance broadly as the traditions and institutions by which authority in a country is exercised. This includes: *i*) the process by which governments are selected, monitored and replaced – the two clusters of indicators used as indices for this are labeled "Voice and Accountability" and "Political Instability and Violence"; *ii*) the capacity of governments to effectively formulate and implement sound policies, reflected by two clusters referred to as "Government Effectiveness" and "Regulatory Burden"; and *iii*) the respect shown by citizens and the state for the institutions that govern economic and social interactions among them – the two clusters labeled "Rule of Law" and "Graft". A preliminary examination of the data in these six clusters shows that Korea does relatively well (compared to the United States and Japan) in the "Voice and Accountability" and "Rule of Law" clusters. It is comparable to Japan in "Regulatory Burden", but has some way to go in "Government Effectiveness", "Political Instability and Violence" and "Graft".

References

Choi, Kang-Shik (1993),
  "Technological Change and Educational Wage Differentials in Korea", *Economic Growth Center Discussion Paper*, No. 698, Yale University.

Kakwani, Nanak, and N. Prescott (1999),
  "Impact of Economic Crisis on Poverty and Inequality in Korea", background paper prepared for the World Bank.

Kaufmann, Daniel, Aart Kraay, and Pablo Zoido-Lobaton (1999),
  "Governance Matters", Washington, DC: World Bank, August.

McKinsey Global Institute (1998),
  Productivity-led Growth in Korea.

Ministry of Commerce, Industry and Energy (Korea) (1999),
  Shaping a New Investment Environment: Korea prepares for the 21st *Century*, September.

Ministry of Finance and Economy (Korea) (2000),
  Korea Economic Update, newsletter of the Ministry of Finance and Economy, 28 February.

National Statistical Office (Korea) (1999),
  Annual Report on the Economically Active Population Survey, Seoul.

OECD (2000),
  Regulatory Reform: Korea, Paris: OECD.

Park, Dong-Hyun (1998),
  "Intellectual Property Regime and the Developing Countries", STEPI, November.

Small and Medium Business Administration (Korea) (1999),
  Survey on SMBs, 27 October to 30 November, Seoul: SMBA.

Thurow, Lester, C. (1999),
  Building Wealth: New Rules for Individuals, Companies, Nations in a Knowledge Based Economy, Harper Collins.

UK Department of Trade and Industry (1998),
  Our Competitive Future: Building the Knowledge-driven Economy, December.

US Department of Commerce (1999),
  "Falling through the Net: Defining the Digital Divide".

World Bank (1999),
  Korea: Establishing a New Foundation for Sustained Growth, Washington, DC: World Bank.

World Bank (2000),
  World Development Indicators 2000, Washington, DC: World Bank.

Chapter 3

Developing Human Resources
for the Knowledge-based Economy

A. Introduction

Increasing overall productivity and adaptability will be fundamental for Korea's transition to an advanced knowledge-based economy. Developing the appropriate human resources will be the linchpin of such a transformation. Ensuring the flexibility of the education system, both formal and informal, will thus be critical for developing creative, knowledgeable and better-skilled citizens for the 21st century.

Korea has attained remarkable achievements in education over the past three decades, and the quality of its basic education has been internationally acclaimed. In recent years, Korea has embarked upon a series of educational reforms. The challenge is whether the well-established Korean education system will be able to maintain its excellence, undertake the necessary reforms and provide the needed human resources for its transition to an advanced knowledge-based economy. To meet this challenge, Korea needs to make its education system more flexible and more relevant to the new global environment by focusing on learning rather than on schooling, creating an enabling environment for promoting creativity, improving the quality of higher education and providing opportunities for lifelong learning. Over-regulation and the continued emphasis on homogeneity throughout the system are observed to be the main obstacles to progress, making a profound and systemic education reform focusing on deregulation and diversification and a new role for government a central requirement. A reformed education system would integrate and update the existing education and training systems, and facilitate learning by anyone, anywhere and at any time.

This chapter provides a brief review of Korea's achievements and recent reforms in education. It then discusses the paradigm shift in education and learning brought about by the new global economy, highlights the challenges facing the Korean education system and proposes possible responses.

B. Educational achievements and recent education reforms in Korea

Korea has a unique education system characterized by much larger private sector representation and investment, and a relatively small publicly financed sector compared to other industrialized nations. In the past three decades, the government has, through its highly regulated and centralized governing system, attained remarkable educational achievements. Since 1970, Korea has maintained full enrolment in primary education (see Figure 3.1). Illiteracy declined from over 10% in 1970 to virtually nil in 1997 (World Bank, 1999a). Gross enrolments in secondary education increased from 40% in 1970 to become almost universal by 1997. At the tertiary level, Korea ranks third among the OECD countries in the educational attainment of its population, and 84% of its high-school graduates entered a university or a college in 1998 (Ministry of Education, 1998). Schools around the country have very low drop-out rates: 0.8% for middle school, 2.1% for high school and 2.5% for higher education. Due to this rapid expansion in schooling, gender disparities have been eliminated at both the primary and secondary levels, although they still exist at the tertiary level (Section D3). At the same time, class sizes have fallen for all levels of education and pupil teacher ratios have become smaller, although they remain considerably higher than the average in certain OECD and other advanced countries (see Figure 3.2 for this ratio in secondary education).

Figure 3.1.   **Education indicators: Korea**

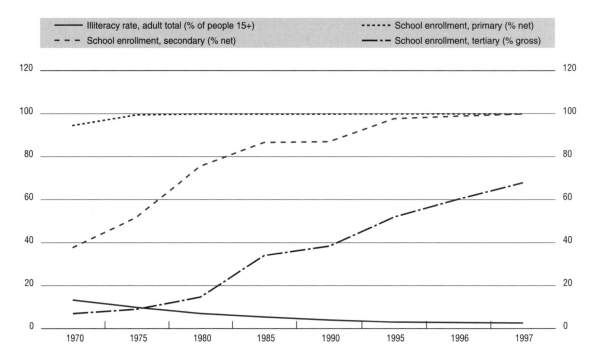

Source: World Bank (1999*a*).

Figure 3.2.   **Student teacher ratio (secondary level)**

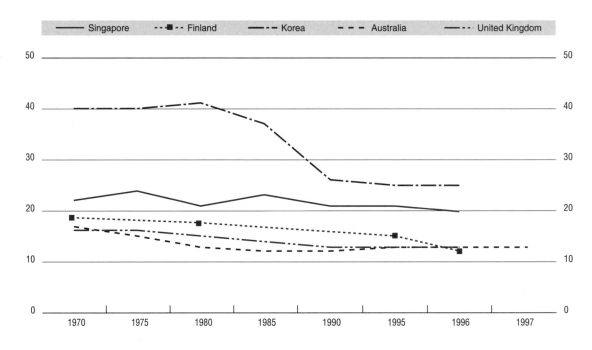

Source: World Bank (1999*a*).

Korea also shows remarkable achievements in other international comparisons. According to the TIMSS study, Korean students at the 4th and 8th grades performed significantly better than the OECD average. In mathematics, for example, Korean students obtained the highest scores among all participating countries, followed by Japan. The transition from school to work is also relatively smooth in Korea – the unemployment rate for young adults (20-24 year-olds) is in the low range among OECD countries. And in tertiary education, although countries like the United States, Australia, and Finland are ahead of Korea with enrolment rates close to 84%, Korea has overtaken Japan and the United Kingdom which had higher tertiary education rates in 1970 but have now fallen behind. Korea's female tertiary enrolment has also surpassed that of Japan and Singapore. In addition, Korea has the highest growth rate in scientific publications among the OECD countries.

Over the past three decades, public expenditure per student as a percentage of GNP per capita in Korea has fallen significantly for tertiary education and increased for primary and secondary education. The total education expenditure increased from 8.8% of GDP in 1966 to 13.3% in 1998 (Korea Education Development Institute, 1998), the highest share for any country at this level of development. Of this, however, only 4.4% was the share of public financing, which is even lower than the OECD average of 4.9% in 1995 (Table 3.1).[1] Such relatively low public investment in education has led parents to spend more on education and, to certain degree, has reduced the strength of the public education system in Korean society.

Parents spend 2.3% of GDP on tuition and school supporting association fees for formal education in both public and private schools from the primary to the tertiary level. In Korea, more than 35% of secondary schools and almost 90% of institutions at the tertiary level are private. In percentage terms, Korean households contribute almost 84% of the costs of tertiary education, significantly more than in any other OECD country. However, lack of choice and large class size in the public sector, combined with a very strict national university admission system, have led parents to invest another 3.2% of GDP on private tutoring for their children. The remaining 3.4% consists of general expenses, including the purchase of study materials, books and stationary, uniforms, transportation fees, lodging, food and other costs).

The Korean Government has embarked upon a series of reforms in education since the early 1990s; these were enumerated in detail in the Presidential Commission on Education Reform (PCER) initiative of 1995, as well as in subsequent revisions.[2] Some 120 tasks for reform have been proposed. Some examples of the reform agenda include: greater autonomy for universities in admissions and academic affairs, including deregulating student admission and enrolment quotas; nurturing leading regional universities to meet the needs of local industry; establishing education and research networks with leading world universities to stimulate research at Korean universities; and realizing the vision of an open educational system through a new Education Credit Bank system. For various reasons, especially the lack of consensus-building in this process, these reforms have not been fully implemented as envisaged.[3]

Table 3.1. **Public expenditure on education as a percentage of GDP, 1995**

Australia	4.5
Greece	3.7
Finland	6.6
Korea	3.6
Norway	6.8
Sweden	6.6
Turkey	2.2
United States	5.0
OECD average	**4.9**

Source: OECD (2000); World Bank (1999*a*).

Given its national zeal for education, Korea has great potential and is well positioned to forge ahead in today's world where knowledge and information are becoming the most critical factors in creating value added. However, a number of serious questions need to be addressed: How can both the quality of learning and opportunities for lifelong learning be improved when the government invests only one-third of the total expenditure on education, and yet controls almost every aspect of its operation? How can educational institutions be more responsive to the needs of the knowledge-based economy while at the same time being constrained by regulations affecting student admission quotas, curriculum, hiring of teachers, and tuition? How can creativity be developed in both teaching and learning when the incentives are such that they overwhelmingly encourage students to pass rigid college entrance exams to obtain admission into the elite schools? Without a massive deregulation and diversification of the existing education system, Korea will be unable to bring about the kinds of changes in the quality of learning required by the knowledge-based economy or the opportunities for lifelong learning that are fundamental for enhancing creativity and productivity in the new economy. The current reform programs do not seem to go far enough in this regard.

C.   The paradigm shift in education

Korea's present education system has responded well to the basic educational needs of the population and was successful in delivering the manpower needed for its industrialization efforts. However, the system was developed to serve the needs of the organisation of production in an industrial society which featured rule-driven, routine production processes controlled by management with little requirement for creativity on the part of workers. The new knowledge-based economy requires a different type of organisation of production, which calls for changes in the relationships of worker to work, worker to worker and worker to consumer, and seeks to stimulate continuous improvements in productivity by giving workers greater control over production (Drucker, 1999). Technological improvements made by workers during the production process are as significant as those resulting from design changes by engineers and scientists. Thus, the knowledge-based economy requires an education system which not only permits learning throughout the lifetime of its citizens and encourages their creativity, but more importantly, is sufficiently flexible to adapt to the changing demands of a knowledge-based economy. Such a system has to be based on continuous learning – from providing basic skills, to developing core skills that encourage creative and critical thinking for problem-solving, to developing specialized skills for specific professional careers and tasks – and should take place in multiple environments: at home, at school and at work. This new environment requires a different approach to work and a different mindset, including not only formal education and the ability to acquire and apply theoretical and analytical knowledge, but above all, a continuous upgrading of skills (Merrill Lynch and Co., 1999).

The concept of "lifelong learning for all", adopted by OECD Education Ministers in January 1996, provides a framework to guide education and training policy in the knowledge-based economy (Hasan, 1998). Successful implementation of this "cradle-to-grave" approach displays four characteristics: individuals are *motivated* to learn on a continuing basis; they are *equipped* with cognitive and other skills to engage in self-directed learning; they have *access to opportunities* for learning throughout life; and they have financial and cultural *incentives* to participate in lifelong learning.

The lifelong learning approach is centered on the learner – defined to include the individual as well as collective entities such as the enterprise, the economy and society at large. The objectives of education policy in this new environment need to be defined in the broader social and economic context and take into account individual objectives as they change over the lifecycle. The economic argument is that the knowledge-based economy requires a skilled, flexible and adaptable labor force. New developments in the labor market include: rising skill thresholds; falling demand for low-skilled workers in relative terms; and rising growth rates for so-called knowledge workers. Micro-level studies indicate increasing use of flexible management practices (OECD, 1998c; OECD, 1999), stemming, in part, from more frequent and volatile changes in the demand for products and services. These developments contribute to reducing the shelf life of skills, and multiple careers through one's working life are becoming more common. All these factors place a high premium on frequent updating of skills, increasing demand for problem solving, teamwork, communication and ICT skills.

In general, based on the accumulated knowledge of international best practices as well as changes in production processes and relationships, the new paradigm in education proposes a new set of elements (McGinn, 1999):

- Primary attention is given to different methods of learning.
- Teachers provide some direction to bodies of knowledge but place greater emphasis on applications.
- Teachers spend little time on instruction and devote the greater part of their effort to the facilitation of learning.
- Students spend most time on experiential learning: learning while working, and working while learning.
- Group learning is emphasized over individual learning.
- Emphasis is on the diversity of perspectives that can be brought to solve problems, rather than on the identification of a single concept or best approach.
- Non-cognitive methods of expression are encouraged to stimulate creativity in finding solutions to "messy" and complex problems, and in facilitating communication.
- Changes in student assessment are required, with a shift away from measurement of knowledge toward evaluation of performance.
- Attention to the application of knowledge increases interdisciplinary work and, over time, leads to a blurring of disciplinary lines and the creation of new disciplines.
- Emphasis in both teaching and learning is on the construction of knowledge through action rather than the (re)discovery of existing facts.
- Different expressions of human intelligence are valued in co-operative efforts at home, in the workplace, in society and in government.

Whether or not everyone agrees with the above set of elements, the emerging evidence from recent education reforms around the world seems to indicate that they are indeed the key drivers of changes in the content, delivery and operation of education systems.

D. Challenges to the Korean education system and possible responses

The new paradigm for education described above, which is characterized by opportunity, creativity and quality of learning, provides a broad context in which to discuss some of the major challenges facing the Korean education system in relation to the knowledge-based economy. Some features of the new paradigm can be identified in the PCER reform programs since 1995 and in subsequent revisions. However, interviews and site-visits in Korea suggest that the main problems identified in the PCER report still persist to a large extent, and that the implementation of the government's reform agenda has not gone as far as originally planned. More assertive reforms in deregulation and administration are needed to create the appropriate enabling environment for educational change. The central government has a very important role to play in the continuous improvements that are needed in the country's education system. However, in the future its role will differ from that played in the past. In this section, the discussion focuses on four interlinked areas which present major challenges to the government and society in the reform of its education and training system: deregulation and decentralization; diversification; gender, relevance and quality; and the provision of lifelong learning.

1. *Deregulation and decentralization*

As mentioned above, Korea currently spends about 13.3% of its GDP on education, and parents spend an additional 3.2% on private tutoring for their children. Such heavy investment in education should, in theory, lead to a satisfactory national education system which provides quality education and meets the demands of the labor market. However, discussions with various stakeholders point to the opposite view: students are less and less content with what they are taught at school; parents are not satisfied with the education their children receive, as evidenced by the amounts spent on extra tutoring; companies complain about the lack of specialty and skills of their newly recruited workers and graduates. The country as a whole is not confident that its education system will enhance its competitiveness in a global economy increasingly driven by knowledge. Why then have the heavy investments in education not produced the results expected by Korean society? Why do Korean parents have to spend so much on private tutoring? In addition to strong pressures from Korean parents for their children to get into elite universities in Seoul, the over-regulated private provision, combined with lack of institutional autonomy and insufficient public funding in the Korean education system, have led to this situation. The time is therefore ripe for a shift in the government's as well as the public's focus away from increasing quantity to enhancing the flexibility of the Korean education system in order to promote creativity and effectively use all the educational resources available to it.

Deregulation. Over the years, the Korean education system, like the country's other economic sectors, has been heavily regulated. On the one hand, such intervention has played a major role in making basic education universal and providing skilled labor for the country's industrialization drive.[4] On the other hand, government intervention has generated a range of regulations regarding the provision of educational services. Although it provides only 4.4% of GDP public spending on education, with two-thirds of its education expenditure coming from private sources and with private education institutions constituting 35 and 90% of schools at the secondary and the tertiary levels respectively, government control over the entire operation of the education system has rendered the system highly centralized and inflexible to market needs. The central government regulates matters such as type of schools, student admission, curriculum, textbooks and teacher hiring; these regulations are also applied to all education institutions, with no distinction between private and public education institutions in the content or type of education services provided. In the absence of deregulation, it will be hard to carry out the educational reforms needed for Korea's transition to the advanced knowledge-based economy.

The most illustrative example of over-regulation comes from the tertiary education sector. Both the private and public universities lack autonomy in their management and academic affairs, with government regulations constraining them in the recruitment and payment of staff, student enrolments and admissions, fee levels, and so on. Although the universities do not receive direct government instructions on curriculum content, the curricula they provide are fairly uniform and many universities simply copy the programs of the top-ranking university. Consequently, the students coming out of this system tend to have the same knowledge structure and skills.[5] Such a heavily regulated system does not have the flexibility to fulfil the country's growing demand for new types of knowledge workers.

Until very recently, the university entrance examination system was open to the criticisms of being both inequitable and inefficient as a means of selecting students, and of causing the entire education system to become exam-driven. Examinations rank students according to a total score in a limited range of subjects, pointing to a lack of diversity; and the adoption of multiple choice questions does not encourage critical thinking or analysis. It is inequitable because it stimulates a high level of spending on tutoring in order to prepare for the entrance examination, putting significant financial pressure on lower-income families.[6] In 1998, the government introduced performance-based evaluation (*i.e.* cumulative school performance in addition to the test scores), and reformed the college entrance system by allowing universities to develop their own admission criteria and select students based on performance rather than on pure test scores. However, this has not yet resulted in substantial changes in the system due to the lack of institutional capacity within the universities to implement such wide-ranging reforms.

The Korean Government has recognized the problems which arise from its tight control over the education system and has embarked on a number of deregulation efforts. Deregulation should focus on: developing an education market with healthy competition; creating flexibility; and promoting synergy in the system to maximize its outcome. Granting greater autonomy to education providers, especially the private institutions, will certainly increase their responsiveness to the needs of the labor market. However, deregulation alone is not sufficient to bring about the required improvements.

Changing the traditional mind-set of the Korean public concerning the pursuit of education, especially at elite universities, is equally important, because such elitism not only dictates the secondary school curricula and the entrance examination system, but also increases the financial burden on parents from private tutoring. Chapter 7 sets out ways to induce such a change in mind-set.

Decentralization. While deregulation may bring the needed flexibility to private institutions, it is not sufficient for public education, which has been subjected to a highly centralized administrative control by the central government. Increased productivity in the new knowledge-based economy will require that decisions be made at the local level, meaning that greater autonomy is devolved to knowledge workers, thus improving their overall performance (Drucker, 1999; McGinn, 1999). Although efforts in the area of decentralization have started, the expertise, knowledge base and supporting infrastructures for effective decision making at the local levels are not yet strong enough to permit the needed decentralization.

Elsewhere in the world, experience shows that the role of central governments in education is changing. This is reflected in the shift of managerial responsibilities and resources away from central government (*i.e.* the Ministry of Education) to lower levels of authority, and often directly to schools, with corresponding accountability. It is hoped that bringing decision-making power and accountability closer to those who teach and manage schools will make them more effective in helping students learn, more efficient in utilizing limited resources, and more directly accountable to students, parents and local communities, thus reducing the need for central control and supervision.

Effective improvement cannot be brought about by schools alone, but it is important that the schools see themselves as a central part of the solution (Fullan, 1999). In Korea, the 1995 PCER reform introduced a new type of governing body in schools, although its roles were not clearly defined. A new autonomous school approach was proposed to improve the quality, competition and diversity of educational services.[7] Still, so far, insufficient administrative and financial decision-making power has been delegated from the provincial educational authorities to individual schools.[8] Further decentralization is therefore needed to enable school-based management and to make the Korean education system more responsive to local needs. Experiences of decentralization in countries like Australia, New Zealand, Spain and the United States can be useful references (see the example of New Zealand in Box 3.1). However, the autonomous school approach is likely to encounter serious opposition as it may be viewed as going against the ethos of Korean culture and as catering only to the higher-income families. The issue of equity that will arise through greater market orientation will also need to be addressed.

Changing role for government. The required efforts in both deregulation and decentralization in the public sector will not entail reduced responsibility for central government in the nation's educational development. Rather, the role of central government will be strengthened, although not in the traditional sense of maintaining control. The central government's new role in education is in macro-level strategic decision making, securing resources, ensuring overall co-ordination and evaluation, setting standards, assuring equity and quality, and providing support services. These new requirements and responsibilities have been recognized in Korea.[9] The role of government in ensuring equity is paramount, particularly in light of the envisaged deregulation of the education system. As the government deregulates and the private sector expands, new demands are placed on government to ensure that poorer students are adequately taken care of through scholarships at the secondary and tertiary levels. The Korean Government has set up a number of programs to help children from poor families avoid the poverty trap: *i)* their school attendance is well protected through a subsidy program; and *ii)* the school lunch program provides meals for those who cannot afford to pay for them.

The Korean Government's share of education expenditure is considerably below the OECD average, and only about 70% of the level in most western industrialized countries. However, as noted above, total educational expenditures in Korea as a share of GDP are probably the highest in the world. Greater deregulation will help the private sector to respond to parents' demand for improved quality and diversity in private education. However, this is likely to be accompanied by rising tuition fees. For children whose parents cannot afford the higher costs of private education, this raises the issue of equity. Therefore, a crucial role for government will be to find ways to reduce the financial burden on

---

Box 3.1. New Zealand's experience with decentralization

From 1987 to 1997, New Zealand undertook a major restructuring of its public education system with the primary objective of decentralizing authority to the school level. With the social pressure (no economic growth, high unemployment yet a lack of skilled people, huge foreign debt and few international trading partners) and political pressure (against the backdrop of broader, sweeping changes in the public sector) in the country, this reform was created by building inter-ministerial co-operation, by employing personal commitment from the Prime Minister to enlist public support, and by providing a clear, "pre-negotiated" roadmap for reform implementation (the Picot Commission Report).

The reform plan was to produce a publicly funded self-managing education system where authority and delivery was decentralized to the local level including: appropriate, well-defined accountability systems; targeting those with specific needs; and, offering choice to all as to where and how to get education throughout their lives. Three key issues were central to this decentralization reform: school-level self-management and decision making; central government setting of clear frameworks, guidelines, standards, and requirements; and parental choice by offering more options, greater diversity, enhanced access. The central government was also committed to targeting resources to needy areas and ensuring equity throughout the system.

While there was a devolution of power to schools, the central government created several intermediary agencies which were accountable to the government, *e.g.* the Education Review Agency, responsible for conducting school-level evaluations; the New Zealand Quality Authority, responsible for quality promotion and assurance; the Implementation/Training Unit, responsible for advising and supporting administrators to deal with the financial management and budget skills; and Education Service Centers for consulting with and supporting schools. Central offices at each level maintained the following roles: policy advisory, managing property, moving funds, and developing personnel, administrative and governance guidelines. In short, their main focus was on equity, support services and results.

In 1989-90, almost overnight, most school-level tasks originally controlled by the central government went to 2 800 parent-elected and -controlled boards. This transition was preceded by the establishment at each school of a Board of Trustees with elected members. These boards signed charters with schools to set guidelines for school self-management and reflect local needs. Many observers feel that they have strengthened the role of the parents, but not necessarily of the teachers.

Some lessons learned from this reform experience are: i) change is catalyzed by a general reform climate, broad public support and requisite infrastructure; ii) consistent, strong and committed political leadership is essential; iii) legislating change demonstrates serious intent and reduces the temptation of reversal by fiat; iv) political leaders must have confidence in the commitment and ability of their public officials; v) allies in the planning and financial arenas of government are crucial, along with other option leaders; vi) use the communications media; vii) change quickly – more quickly than comfortable; viii) appoint a specialist change manager with authority to act; ix) take action to gain the co-operation of existing staff who must implement the transition; x) deliberately engage or isolate pressure groups – never ignore them; xi) be prepared to inject extra funding to "sweeten the pill" for reform losers; xii) allow for mistakes and create incentives for transparency and learning; and xiii) remember exit, transition and entry arrangements. This decentralization reform has provided more choices for parents and has improved the effectiveness and efficiency of the public education system in the country. Although there are some indications of the increased learning outcome of students, thorough evaluations are currently under way.

Source: Perris (1998).

---

parents looking for quality education for their children, while maintaining equity in the system through the use of instruments such as scholarships, vouchers, tax deductions, low interest loans for private education and improvements in the quality of public education. It should closely examine the efficiency of the allocation of public and private resources to education. In addition, the government needs to provide substantial support to help individual institutions build their capacities. To promote quality and accountability, the central government should make the system more transparent and increase the availability of educational policy information and statistics to the public. Without this, deregulation and decentralization efforts will not succeed.

Possible responses

- Continued deregulation of the education system is needed to further educational reform in Korea. This includes deregulation not only of the tertiary, but also of the secondary education system.

- Greater efforts should be made to introduce a market-oriented and stakeholder-driven governing system in education in Korea, with clearly defined autonomy and accountability at the institutional level.

- Responsibilities should be decentralized, with corresponding resource allocations, and with each level being held accountable for results, especially to the local level. Institutional autonomy needs to be increased to enhance the decision making of schools[10] and universities[11] in areas such as staff administration (hiring, firing, salaries and benefits), teaching loads, admissions policies, enrolment quotas, and budgeting and financing.[12]

- The government should continue to increase public expenditure and investments in education, enabling the system to provide high-quality services and reduce the financial burden on parents of private tutoring. It must ensure equity in the system by investing in appropriate schemes, for example, providing scholarships to students from poorer families.

- With the increasing diversity in the supply of education, it will be important to develop a transparent mechanism for the accreditation of different types of institutions and for undertaking benchmarking assessments to judge the quality of teachers and programs and to track and monitor quality over time. A national body with clearly defined criteria and standards could be established to undertake this task. The government has taken a number of steps in this area in recent years, such as the establishment of an educational credit bank system which allows individuals to accumulate course credits from accredited institutions toward certification throughout their lifetime. Further efforts need to be made along this path to set up a national framework that encompasses all educational and vocational qualifications, whether obtained formally or informally.

- The newly conceived college entrance system, which allows each university or college to develop its own admission criteria, needs to be firmly implemented. This will improve equity of access to university education, and help to shift the learning culture in high schools away from learning by rote toward the development of critical-thinking and problem-solving capacities. The government needs to develop instruments such as scholarships and voucher schemes to guarantee equal learning opportunities to all Koreans.

2. Diversification

As the knowledge-based economy evolves and the Korean education system is further deregulated, there will be increasing demand for diversified learning opportunities and increasing competition among providers to provide a varied menu of offerings.[13] A diversified system would consist of centers of excellence, providing leadership for the nation's education and training system, and a variety of quality institutions which would form the needed critical mass to serve the needs of both the local economy and individual learning. This would involve not only universities and vocational institutions, but general secondary schools as well. Consequently, their programs would permit some diversification, while maintaining common core standards set by the Ministry of Education. In addition, increased possibilities for self-paced and directed learning – brought about by the utilization of ICTs and by a blurring of the distinction between formal and informal institutions in terms of services provided – require greater diversification in both the content and the providers of education and training.

Diversification of the education system has already started in Korea, as for example, in the government's *Seventh Curriculum Reform* [14] and *Brain Korea* 21 initiatives which aim at establishing a few centers of excellence in higher education with substantial public investment. However, investing in centers of excellence alone will not satisfy the demand for human resources with the diverse skills that are needed for a knowledge-based economy: some structural adjustments are also needed. For instance, the scale of junior colleges is currently only about the half of the formal colleges and universities, even though junior colleges produce the middle-level technicians that are increasingly in demand in the new economy. At both the college and school levels, very few programs are challenging enough for gifted students whose talents are important assets for the development of the knowledge-based economy in Korea. Thus, the kind of diversification in education and training needed in Korea is not limited to any particular level of education or institution; rather it calls for a systematic effort.

Currently in Korea, the high-school vocational and academic streams are completely separate and opportunities for constructing individual pathways combining elements of academic and work-related experience are limited. This implies that the most important educational decision in a child's educational history is made at age 15, at the end of the junior school, and is based mainly on examination scores. Such early streaming constrains individual learning opportunities. Recently, several academic high schools have begun to offer the vocational program to their senior students, although this has met with considerable resistance from students and parents, who worry that it might lower their score in the annual nation-wide Scholastic Achievement Test (SAT) administered by the government.[15] In addition to the diversification in curriculum, other efforts should include, for instance, magnet schools in technical areas or in the arts, as well as specialized training institutions outside the formal school system.

Since tertiary education is the main provider of highly skilled personnel, Korea needs to strive to promote diversity in its tertiary education system. Although the nearly universal participation in tertiary education seemingly provides enough knowledge workers for the Korean economy, there is a growing need for a more diverse body of knowledge workers. Under the government's rigid regulations, the Korean universities operate in a "department store" style, offering similar disciplines rather than concentrating on selected areas in which they have a competitive advantage. Recently, the government has started to expand the autonomy of the universities, particularly in creating their own curriculum. However, take-up has been slow due to the lack of institutional capacity in many universities and the lack of competition among them. A stronger competitive environment needs to be cultivated, as this will encourage the universities to offer greater choice and become more responsive to student needs.

The current higher education system in Korea is concentrated around Seoul and its vicinity, and there is a strong concentration of the "best" universities in Seoul. This has given rise to a strong sense of elitism centered on Seoul National University (SNU). It also inspires a disproportional share of high-school graduates to apply for entry to SNU and other elite universities in Seoul at the expense of provincial universities. Provincial universities, in turn, tend to follow the trends set by SNU, often without regard to their ability to afford them. The result is a tendency towards homogeneity in public universities that reduces the flexibility and relevance of provincial universities in addressing local development needs. There is a need to further develop local universities outside Seoul. Graduate schools in different regions need to become specialized to meet local needs. It is important to note that diversification of the tertiary education system will involve complementary changes at the secondary level. The strong belief of the Korean people in education, and especially in elite education, and their reluctance to experiment in this area, means that diversifying the national educational system will present quite a challenge.

Possible responses

- Measures to increase institutional and curricular diversification should be combined to promote interdisciplinary studies and advance broader qualification profiles and flexibility, including overlap between academic and vocational streams. It is important to ensure students' individual choice and to provide for mobility within the system to fulfill the goals of lifelong learning, equity and efficiency. This could, for example, be done by institutionalizing the credit transfer system; accepting double majors; providing more information on other universities and colleges; providing economic incentives for mobility; and facilitating exchange programs with national and foreign universities.

- Greater emphasis should be given to developing "double qualifying educational pathways" that can lead students to university or to tertiary-level vocational education or directly to the labor market. The axis of vocational education should move to the post-secondary level to meet the demands of the knowledge-based economy. Vocational colleges should be authorized to teach and award degrees and their degree programs should be given proper recognition in the tertiary education system.

- Efforts should be made to develop a diverse tertiary education system with a few comprehensive universities that cover all disciplinary areas and others that provide more specialized programs. Each institution needs to develop its own programs in light of its institutional strengths and local industry needs. As virtual universities mushroom around the world, diversification could be achieved by using the new technologies. In addition to the emphasis on universities, efforts should be made to upgrade and expand junior colleges. Secondary-level education should be diversified through the deregulation of the education system.

3. *Gender, relevance and quality*

Gender. Korea's successful transition to a knowledge-based economy needs to be underpinned by a positive attitude to reform and development, especially in the area of education. A key issue in this area is the gender bias against women which is deeply embedded in Korean culture and tradition. For example, although female students make up approximately 38% of total enrollment in tertiary education, they are mostly concentrated in the arts and social sciences. Only 4% of graduates in science and technology are women. Female university professors are rare, particularly in the science and technology fields. Lesser participation of women in S&T limits the pool of women available for recruitment in the technology industries. Gender biases in textbooks, prevailing social norms and a strongly gender-differentiated workforce make it difficult to change deeply entrenched attitudes to women's employment.

The change in attitude toward the gender issue needs to start at an early stage. The curriculum at schools needs to be re-examined to remove gender bias and to take account of gender issues. In addition, women should be encouraged and given the opportunities to assume leadership roles in the education system. This change of attitude toward women in society will be a long-term process, but it is crucial if each Korean citizen is to fully participate in and contribute his/her full potential to the nation's transition to a knowledge-based economy.

Relevance. According to Robert Reich, the three jobs of the future include routine production services (performing repetitive tasks); in-person services (providing person-to-person services); and symbolic analytical services (problem-solving, problem-identifying and strategic brokering activities).[16] In Korea, the overarching need is to develop critical-thinking skills, communication skills, computer skills, quantitative reasoning and social interaction among students so that they can become successful knowledge workers for the future.[17] These new skills are very different from those traditionally taught at schools. Although the quality of Korean education at the primary and secondary levels has been praised around the world, how well can Korean schools perform and how good will be the quality of Korean education when measured against these new skill requirements? Until now, the main purpose of secondary education in Korea has been to prepare students to enter prestigious colleges and universities. Even though a number of reforms have been instigated in this area,[18] the college entrance examination system still tends to give priority to students who do better in all fields on average, rather than to those who possess creativity and problem-solving capabilities. The gap between what currently exists and what is required for the future needs to be bridged for Korea's successful transition to a knowledge-based economy. There are distortions in the labor market in Korea with respect to wage premiums for different types of education/degrees, partly fueled by the traditional sense of educational elitism. In its transformation to an advanced, knowledge-based economy, Korea must analyze its education system – not only from the supply side, but also from the demand side.

Korea has been attempting to achieve the right balance between quantitative expansion and quality promotion of its university education system. A shift in Korean education policy in 1980 increased the number of university entrants by 30%, and reduced the wages of young college graduates relative to the wages of young high-school graduates between 1982-88 (Choi, 1999). Firms complain that they now have to retrain newly recruited college graduates, although Korean firms spend much less on such training than do enterprises in other OECD countries. Korean society needs to realize that the knowledge-based economy calls for increased technical skills and that this does not necessarily require a university-level education.[19] This links back to the diversification issue.

Korean universities do not maintain close linkages with industry and there are no mechanisms for systematic feedback from industry to indicate the kind of knowledge that should be taught and the skills that should be gained in universities. As a result, the university curriculum does not reflect the needs of industry, and graduates are not well prepared to perform their jobs without further training. A better mechanism for integrating the demands of the market into the education system is required if Korea is to transform itself into a knowledge-based economy. The KBE calls for university faculty to become "learning facilitators" and "knowledge utilization co-ordinators", training students and transforming research results into marketable products (Gibbons, 1998).

Quality. The quality of the basic education system in Korea has been acclaimed the world over and Korean students do remarkably well in international comparisons. The question is whether the system is of sufficiently high quality to ensure the development of the Korean knowledge-based economy, especially in terms of the new skills required (Murnane and Levy, 1997).

Teachers play an essential role in ensuring the quality of education, but the lack of a sound incentive system for teachers is a major problem facing Korean schools. The salary rates of beginner teachers may not be sufficiently competitive with starting rates for graduates recruited to other occupations. In the area of teacher education, dated curricula and pedagogy are some of the other major problems and have contributed to the difficulties encountered in recruiting high-caliber people to the teaching profession. This is a problem as teachers are increasingly seen as fundamental agents for change and quality assurance in many countries which are undertaking major education reforms (Fullan, 1999; Darling-Hammond, 1997; Elmore, 1996). It is particularly true that while decentralization has brought decision making and resources closer to those who teach and manage schools, it simultaneously demands schools and teachers to be more accountable to their students, parents and local communities. The increased accountability and requirements for new teaching skills call for a strengthening of the teaching profession; but the teaching profession itself must also become more active about its professionalism (Chase, 1998).

Korea lags behind in this area: in 1995, 55% of all secondary teachers came from universities that did not specialize in teacher education. Teachers are recruited at the local level on the basis of selection tests administered by district education authorities, and undergraduate students focus narrowly on preparing for these quite mechanistic tests, rather than securing a broad knowledge and pedagogical basis for teaching. Teaching practice (in all eight weeks for primary and four weeks for secondary) and pedagogical training play a remarkably minor and insufficient role in teacher education and training. The curricula for teachers' education gives far too much weight to subject content and too little to cross-subject themes such as those emphasized in the KBE – communication, problem-solving, critical thinking, etc. The curricula is also dominated by university academics who do not promote teaching methods that focus on the acquisition of knowledge through practical experience. In addition, in-service education and training seems to be restricted to teachers at certain promotion points in their careers, instead of being related functionally to the changing requirements of the teaching profession as a whole. Insufficient skills and training in technology have discouraged teachers from integrating ICT with curriculum and teaching. Experience and lessons can be learned from countries such as Singapore where, based on the national ICT strategy, a teacher training plan has been developed with different approaches (*e.g.* a formal support system, informal networks of teachers and self-directed independent training) focusing on IT integration at the academic subject level. In short, to improve the quality of its education system, Korea must establish appropriate incentives and mechanisms for the professional development of teachers. The government has an important role to play in this area, not only in terms of providing the needed supporting services, both financial and non-financial, but also in making the current system more transparent to enhance its quality.

Quality is a more challenging issue in Korea's higher education system. Korean universities tend to focus on teaching students subject matter content ("know-what" knowledge) rather than teaching or coaching them on how to apply or utilize knowledge ("know-how" knowledge) (Foray and Lundvall,

1996). Lack of a sound incentive structure, including unsophisticated evaluation measures for tenured professors (*i.e.* which are seniority-based rather than performance- and outcome-based) and hiring own graduates, has been a major contributor to the low quality of faculty. This, in turn, makes it more difficult to address quality issues at the institutional level. Once hired, professors have lifetime tenure irrespective of research/teaching performance, their pay is linked with the civil service scale and promotions are based on seniority and length of tenure. Universities typically also hire their own graduates rather than searching out the best performers irrespective of the candidates' alma mater.[20] Unless the incentive structure is reformed, no more than minimal gains may be expected. Thus, unlike its basic education system with its highly recognized quality, Korea has neglected to promote quality education in its university system, which is vital today in building a strong knowledge-based economy.

In addition, Korean universities have weak linkages with the international educational community. Students do go abroad for education but, due to government-mandated criteria for setting up campuses, foreign teaching institutions have not been given incentives to provide educational services in Korea as they have in, say, Malaysia and Singapore. These criteria include size of the campus lot, quantity and quality of various facilities, size of buildings, number of professors, etc. Foreign schools have difficulty in meeting these regulations due to the very high cost of space and facilities in Korea, particularly in Seoul. There are also informal pressures by students and the government to keep tuition fees low for all Korean universities. This is a missed opportunity in terms of raising the quality of Korea's higher education. The government could, for example, start by creating incentives to encourage joint offering of degrees by foreign and domestic universities. Under this format, foreign universities would maintain linkages to operations in their home countries. For example, students could study for part of the time in Korea and part of the time at their home campuses.

One of the main barriers for foreign faculty teaching in Korean universities is the low English-language proficiency of the students. The current government has introduced several measures to improve the quality of university education in Korea. Under *Brain Korea* 21, the government plans to invest USD 1.2 billion over the next seven years in preparing the Korean workforce for the coming century. However, attaining standards in higher education which are comparable to those of leading countries is not simply a matter of resources. Many fundamental changes will have to be made at the institutional level regarding organisational structure, curriculum and pedagogy to match the requirements imposed by globalization and competition.

Possible responses

- Strategies to increase the number of women who can participate fully in the KBE should include the removal of socio-cultural obstacles to women who are skilled and have technical expertise. This should include the desegregation of academic fields by gender, and the provision of support, networks and incentives to encourage women to enter, remain in and succeed in areas of study that will enable them to work in any field. At school level, teachers should be educated about gender biases and expectations; curricula and textbooks should be adapted to remove gender biases; and increased career guidance should be available for girls. At tertiary level, women should be encouraged to take up science and technology courses through: putting in place outreach measures such as grants and scholarships; providing support networks for women who are in large minorities on courses; setting quotas for women for lectureships, post-graduate courses, research positions; and ensuring equal opportunities for overseas training.

- There is a need to reform the existing curriculum and teaching pedagogy to make learning experience gained at school more relevant to society's expectations. The integrated curriculum should include training in the new skills, *e.g.* communication skills, capability to utilize ICTs, problem-solving capabilities, social skills to co-operate and share with others, etc.

- Efforts should be made to revise the curricula to improve the balance between practical and theoretical subjects, and incorporate new technologies. New technologies will need to be an integrated part of teaching methods in most subjects in order to secure skilled and up-to-date graduates. This will require general access to computers, the use of computer-based training, access to the use of electronic mailing, distance education, and so on.

- Greater importance should be given to developing more extensive and varied forms of university-industry partnership.[21] Such improvements would lead to a more effective, efficient university system – one which is more relevant to the needs of industry and which prepares students to be more productive on entry to employment. Measures which foster the establishment of exchange arrangements between universities and enterprises should also be encouraged.

- To improve their educational services, universities could consider the adoption of new instruments including: i) accepting double majors; ii) adopting more open-ended evaluation and testing methods; iii) facilitating student and teacher mobility both nationally and internationally; iv) integrating field experience in the curricula, such as compulsory traineeships in private enterprises; and v) formalizing credit transfers among institutions.

- There is need to reform the incentive system for teachers, at both the school as well as the university level, by designing pay schemes that are not based on seniority, but rather on performance and outcomes.

- It is also important to adopt effective mechanisms to encourage healthy competition among professors and teachers.[22] A knowledge management system could be introduced to develop communities of practice aimed at upgrading the skills of teachers to teach with new technologies and pedagogy, through sharing of experience and best practice.

- Another important step to improving the quality of education in the country will be to encourage the expansion of exchange programs between Korean and foreign universities in OECD countries. Foreign professors could be appointed in Korean universities to help them keep up with the latest developments in their subject areas. To stimulate competition among Korean institutions, the government could consider gradually opening the education market to accredited foreign providers in tertiary education.

4. Lifelong learning

In the context of the rapid development and dissemination of new knowledge, it is particularly important to facilitate opportunities for lifelong learning. It should be emphasized that lifelong learning is a "cradle-to-grave" concept and is not limited to adult learning only. In advanced countries, a proliferation of institutions and mechanisms focus on lifelong learning at work and at home. New forms of learning are becoming available, such as the Internet, increased use of TV, distance learning, virtual universities, and others. These trends can only intensify in the knowledge-based economy.

The lifelong education system is not well developed in Korea. ICTs are not fully exploited in areas such as distance learning and virtual universities, and Korea lags its OECD peers in the provision of learning opportunities for adults. While the training opportunities provided by the private sector are limited in comparison to other advanced countries, those that are available could lead to a risk of social polarization, as those who are better qualified may have the most opportunities for further training. However, there will be no incentive and motivation for lifelong learning if the improved or newly learned skills are not appreciated in the workplace. Such recognition and reward are important because in Korean culture, rewards are often based on seniority rather than on increased productivity derived from improvements in workers' skills.

Since learning needs and motivations change over the lifetime, the knowledge-based economy needs to cater to the growing diversity of learning needs, both formal and informal. This is important for developing informed citizenship, strengthening the foundations of democracy and developing a consensus for the KBE. Thus, the issue of access to knowledge and information for all, and especially for the poor, needs to be stressed.

The Korean Government has recognized the importance of this issue and has begun to expand the high-speed Internet network, offer free public access to computers, provide training in informatics and promote English proficiency. With regard to the digital divide for the children of poor families, the government has set up a special program to teach Internet and basic PC skills to nearly half a million students from poor families. It will also give free PCs and five years of free Internet access to some 50 000 children from low-income families (see also Chapter 4 of this report) (Kim, 2000). These measures are important, but not sufficient. For example, teaching children how to use the Internet is only the first step. The Korean Government should adopt some of the more efficient solutions that have been adopted by many countries, both developed and developing ones. The United Kingdom plans to set up 700 local Internet access centers around the country to increase access to knowledge and information. Canada has already connected all its libraries and schools. However, current programs to help poor students and other measures such as IT-related local government programs to build the information society in local areas, are not closely integrated in Korea as the main agencies dealing with these issues are different and effective co-ordination between them is lacking.

Another important issue is the development of content for self-learning students. For example, Canada (SchoolNET) has put in place incentive programs to encourage teachers, professors and students to develop content that can be shared.[23] Korea needs to develop suitable content for students at primary and secondary schools, and to make use of content developed by other countries.

Possible responses

- Korea needs to fully integrate its current adult, informal, vocational and distance education systems with its formal education system, and to develop a lifelong learning system. The Education Credit Bank system is one initiative that can help to ensure such integration.

- The government should invest more resources into helping schools build up their information infrastructures,[24] and facilitating greater inter-university networking through ICTs to improve teaching and research.

- The development and expansion of early childhood educational services should be given a higher priority in the government's educational agenda (including funding). A closer integration of early childhood educational services with the first year of primary education should be undertaken.

- Women should be targeted in policies for enabling lifelong learning. Sites for early childhood education can help women to access second-chance educational opportunities. The government should provide guidance and information services about educational programs and training, introduce support groups (as in the case of Singapore), and provide access to open education programs and learning networks.

- ICT-supported network learning should be utilized more in the education system to overcome the limits of distance and time.[25] Special attention needs to be given to the integration of ICTs with curriculum and pedagogy. The government's efforts toward equal access should gradually enable every citizen to learn whenever and wherever he/she wants in a convenient way and at low cost.

E. Conclusions

The planned transition to a knowledge-based economy requires that Korea undertake systemic reforms in education. These reforms will involve profound changes in all aspects of its current education system, *i.e.* not only in the inputs and their content and composition, but also in the institutional and governance infrastructures that deliver them. Deregulation and diversification of the existing education system is fundamental as, in its absence, the system will remain inflexible, lack competition and be less relevant to the needs of a changing society. At the same time, the education system must respond to the needs of the population and should emphasize lifelong learning. It should also be noted that the education reform process is complex and will take time – there is no instant solution guaranteeing quick results.

Global experience and research on education reform show that factors such as shared vision, clear goals, strong leadership, broad consensus building, the ability to make adjustment along the process, measurable indicators and the availability of a support infrastructure are critical to ensuring that the reforms achieve the desired results. The government will play a central role in this process: overseeing the whole education system, particularly with a view to maintaining minimum standards, setting up a system of national qualifications, providing information on outcomes and quality, improving teacher training, encouraging industry-university partnerships, dealing with issues of equity and investing more in public education.

In short, the systemic reform in education required in Korea will be a long-term process.[26] The challenge is how to maintain the good practices achieved by the existing system, such as relatively high average quality and low dispersion of student performance, while creating a new system that will better serve the country's human resource needs in the global economy.

It is suggested that Korea study how other countries, especially the OECD countries and neighboring countries like Singapore and Hong Kong, are undertaking systemic reforms in education, with special attention to process-related issues. Above all, a change is needed in the mind-set of major players involved in decision making on the acquisition and value of education, training and skills. The required changes need the championship of the government and extensive participation of all stakeholders: parents, teachers, unions, private sector, professors, other members of civil society and the media. The Korean Government has demonstrated its strong commitment to education reform. Given the capacity and the political will to carry out such unprecedented reforms, Korea will benefit from a new generation of skilled citizens and will be in a strong position to make the transition to an advanced, knowledge-based economy. Chapter 7 elaborates on the implementation of the reform of the Korean education system.

Notes

1. In 1995, in terms of public expenditure on education as a percentage of GDP, Korea was in the lowest group among the OECD countries, together with Turkey and Greece. Korea's expenditures were only about half of those of the Scandinavian countries (OECD, 1998a).

2. Although the PCER has since been dissolved, the proposals highlighted are in various stages of implementation.

3. In 1998, the Korean Ministry of Education launched a new campaign to create a "Five Year Education Development Action Plan (draft)". The latest reform package, entitled "Vision for Education beyond 2002: Creation of a New School Culture", comprises five components: creating an autonomous school community; implementing a student-centered curriculum; cherishing the value of students' life experiences; diversifying the methods used for evaluating students; and emphasizing the professionalization of teachers. Detailed programs for each of these components are under development.

4. Although it has been argued that the educational policies pursued by the government have artificially stimulated the public's demand to seek higher education, one cannot ignore the positive role played by the Korean Government in the quantitative expansion of schooling in the country (Ki Su Kim, 1999).

5. Woo and Lee (1999). The PCER reform proposals include increasing institutional autonomy, and implementation has already started, for example, in the regulation of student admissions to universities. However, the reforms do not go far enough.

6. Parents spend large amounts on private tutoring, some to make up for poor-quality schooling, and to provide greater diversity. More than 53% of Korean households with school-age children pay for private tutoring for their children (KEDI survey). More than 50% of students receive tutoring from the for-profit Private Learning Institutes (PLI) or Hak-Won. Total expenditures on private tutoring amounted to 3.2% of GDP in 1998. There are two kinds of private tutoring (PT): one for improving talents and the other for academic achievement. Typically, pre-school children and elementary school students engage in private tutoring for the purpose of acquiring specific talents, such as drawing or music, and for improving the 3Rs (reading, writing, arithmetic). This may be called PT1. This type of tuition is demanded because Kindergarten or primary schools do not meet parents' expectations of quality and variety in education. Secondary school students, on the other hand, focus on academic achievement in such subjects as English, Korean and Math in order to improve their chances for college entrance. Expenditure on PT2, therefore, does not so much compensate for poor school quality but is geared largely towards the college-entrance exam, which is highly competitive because of the limited university entrance quota.

7. This approach advocates, for example, freedom to charge tuition fees at private schools and freedom to set curricula at all schools.

8. For example, an examination of the percentage of decisions taken at each level of government in public lower secondary education in Korea in 1998 shows that 37% of decisions were taken at the central level; 31% at the provincial/regional level; 7% at the sub-regional level; none at the local level; and 25% at the school level. In contrast, in some Scandinavian countries (Finland and Norway) and the United States, more than 50% of decisions were taken at the local level (OECD, 1998a).

9. The Korean Government provides the framework for the curriculum, qualifications, quality assurance procedures and standards for teacher training. Within this framework, schools have substantial responsibility for managing their own affairs, including budgets and staffing. This has been the trend in many countries around the world (e.g. Australia), informed by research from organisation theory and school effectiveness. It reflects developments in other sectors, including business, and is likely to be sustained. The school remains accountable to a central authority for the manner in which resources are allocated, usually judged to be effective (or not) by student outcomes.

10. As regards the autonomous school approach, one way may be to introduce such a scheme on a pilot basis, while addressing the equity issue by giving access to all students within an area (or through a lottery approach), their tuition being paid through some form of income-based subsidies, e.g. a combination of scholarships and vouchers. The purpose of such a pilot would be to demonstrate that school-level initiatives can improve the quality of education.

11. Management capacity at the schools and university level should be strengthened. Increasing autonomy might require changes in the basis of allocating public funds to universities, by for example, moving towards block grants to take account of the need for local decision making and management, and to increase effectiveness and efficiency.

12. The aim would be to permit institutions to attract quality staff in a competitive market, provide them with competitive employment conditions (academic as well as financial), adjust staffing and resources in line with changing demand for courses, allocate resources between teaching and research, etc. The corollary of most grants of autonomy is that institutions become more financially self-sufficient.

13. In addition to the need for people who are creative and sufficiently flexible to meet a variety of demands, there will be increasing demand for a very large number of knowledge workers ("technologists" in KBE jargon). These workers – who may constitute the fastest-growing group of all workers – do both knowledge work and manual work, and, for example, include health-care workers and automobile mechanics. In the transition to the knowledge-based economy, more and more manual workers will be technologists. In increasing knowledge-worker productivity, augmenting the productivity of technologists deserves to be given high priority (Drucker, 1999). The training of technologists is weak in the current Korean education system. For example, a recent firm-level survey for Korea indicated that the quality and supply of technicians and high-skilled personnel poses a problem in production and for technological improvement (World Bank, 1999b). However, educating technologists is expensive and calls for partnerships with all sectors of the economy. Another issue is that in Korean culture and social structure, educated people look down on people who work with their hands.

14. The Seventh Curriculum Reform offers a number of reforms, including a reduction in core subjects and a greater degree of choice. At present only limited changes in majors (20%) is permitted. Special admissions are allowed only up to a total of 10%. This type of diversity is linked to the freedom of universities to control the size and nature of their enrolment.

15. The number of students in vocational high schools was about 42% in 1995, with the remainder in academic high schools. This percentage is among the middle ranks of OECD countries.

16. Examples of routine production services include traditional blue-collar jobs and low- and mid-level managers (25% in the United States); examples of in-person services include waiters and flight attendants (30% in the United States); and examples of symbolic analytical services include scientists and lawyers (20% in the United States) (Reich, 1991).

17. For example, it has been said that the minimum skills now needed to obtain a middle-class job include the ability to: read at the 9th-grade level or higher; do math at the 9th-grade level or higher; solve semi-structured problems where hypotheses must be formed and tested; work in groups with persons of various backgrounds; communicate effectively, both orally and in writing; and use personal computers to carry out simple tasks like word processing (Murnane and Levy, 1997).

18. Some significant liberalization has taken place in the area of college admissions. Starting in 2002, universities will have complete freedom to use their own criteria in selecting students.

19. For example, in the State of Massachusetts, more than 70% of high-school graduates receive some form of tertiary education. There is constant pressure on schools to increase offerings in science and mathematics, and there is declining support for vocational education and training. This one-sided approach to education ignores the reality of how economies and labor markets function. According to the State's estimate, between 1996 and 2006, the economy will create through job replacement and growth about 1.1 million job openings. Only 3% of these openings will require schooling beyond university; 31% will require at least a university education; more than 54% of the jobs will be in occupations that require high school or less; and 36% will be in occupations that require no more than one month of training (US Division of Employment and Training, 1998).

20. The Ministry of Education has recognized these problems and revised the rules of the Public Educational Personnel and Staff Act. Under the new rules, contract-based faculty hiring is promoted, quotas are set to limit own-graduate recruitment by universities, and the universities will decide their own standards for evaluating their faculty.

21. Close links between university and enterprises can be very profitable for universities, since they can secure up-to-date and relevant teaching at the universities. On the other hand, industry will often also be able to profit from research results generated by universities. Empirically it seems that a close co-operation between universities and enterprises facilitates development and innovation.

22. In the past, the government emphasized equality in access to education; as a result, competition was not given much importance, and professors and teachers were treated equally regardless of the difference in their performance. This distorted attitude has had a negative influence on the adoption of a fair evaluation system for teaching staff.

23. For further information, see *http://www.schoolnet.ca*.

24. The average number of students per computers was 16 in 1998. Vocational schools showed the highest PC distribution (5.5), while academic high schools showed the lowest PC availability rate (19.6).

25. For example, the United Kingdom's National Grid for Learning scheme highlights the UK Government's commitment to lifelong learning and the creation of a learning society. The main aims of this scheme are to help create a "connected society", to improve the quality and availability of educational materials and to widen access to learning.

26. For a summary of lessons learnt from the Korean education reform effort, see Park (1999). The main points are: i) for the success of any reform, close co-operation, together with a system of checks and balances between the Commission (designing body) and the Ministry of Education (implementing body), is important; ii) sufficient budgets should be allocated for the reform; iii) in dealing with vested interest groups, more refined, sophisticated and flexible strategies should be developed, since merely appealing to the moral conscience, civic virtue or public-mindedness of the interest group will not work; iv) it is important to provide the appropriate incentive system to bureaucrats; v) a special pay system and promotion policy could be designed to motivate them to be more proactive for reform; vi) for effective deregulation, the establishment of a "Committee for Deregulation" composed only of non-bureaucrats is recommended; vii) close co-operation with the media is indispensable for the success of the reform program; viii) for reform to succeed, a top-down (government-led) approach should be accompanied by a bottom-up movement; ix) participation of teachers and parents in the design and implementation of the reform process is of central importance in institutionalizing their permanent participation in educational policy making and in the implementation process; x) as educational reform usually takes a long time to consolidate, the reform group should prepare a series of smaller victories or reforms that can be celebrated along the road, thus enabling a high level of enthusiasm for reform to be maintained; and xi) the best way to successfully consolidate reform is to set up a reform body that outlives the current government or regime. This will ensure consistency and continuity in a politically and institutionally guaranteed reform management process, thus leading to successful consolidation of the reforms.

References

Chase, R. (1998),
"NEA's Role: Cultivating Teacher Professionalism", *Educational Leadership*, Vol. 55, February.

Choi, Kang-Shik (1999),
"The Impact of Shifts in Supply of College Graduates: Repercussion of Educational Reform in Korea", *Economics of Education Review*, Vol. 15, No. 1.

Cole, George (1999),
"Great Leap Forward", *Times Educational Supplement*, 15 October.

Darling-Hammond, Linda (1997),
"Reward and Reform: Creating Educational Incentives that Work", in S. Fuhrman and J. O'Day (eds.), *Restructuring Schools for High Performance*, San Francisco: Jossey-Bass.

Drucker, Peter (1994),
"The Age of Social Transformation", *Atlantic Monthly*, November.

Drucker, Peter (1999),
"Knowledge Worker Productivity: The Biggest Challenge", *California Management Review*, Winter.

Elmore, Richard (1996),
"Getting to Scale with Good Educational Practice", *Harvard Educational Review*, Vol. 66, No. 1.

Foray, Dominique, and Bengt-Åke Lundvall (1996),
"The Knowledge-based Economy: From the Economics of Knowledge to the Learning Economy", in OECD (1996), *Employment and Growth in the Knowledge-based Economy*, Paris: OECD.

Fullan, Michael (1999),
"The Return of Large-scale Reform", *Journal of Educational Change*, October.

Gibbons, Michael (1998),
"Higher Education Relevance in the 21st Century", Washington, DC: Education, Human Development Network, World Bank.

Hasan, A. (1998),
"Lifelong Learning: A Monitoring Framework and Trends in Participation", in OECD (1998b), *Education Policy Analysis*.

Kim, Ki Su (1999),
"A Statist Political Economy and High Demand for Education in South Korea", *Education Policy Analysis Archives*, Vol. 7, No. 19.

Kim, Sungteak (1999),
"Policy Suggestions for the Development of Knowledge-based Industry in Korea", KIET, June.

Kim, Eun Jeong (2000),
"Closing the Digital Divide in Korea", e-mail received on 6 April, Washington, DC: World Bank.

Korea Education Development Institute (1998),
Study on Education Expenditures in Korea, December.

McGinn, Noel (1999),
"Globalization, Education and Knowledge for Development", speech given at the Workshop on "Using Knowledge for Development", Helsinki, Finland, 25 May.

Merrill Lynch and Co. (1999),
"The Book of Knowledge: Investing in the Growing Education and Training Industry", San Francisco.

Ministry of Education (Korea) (1998),
Educational Yearbook of 1998, Seoul.

Ministry of Education (Korea) (1999),
Brain Korea 21, Seoul.

Ministry of Finance and Economy (Korea) (1999),
"A Comprehensive Plan for Knowledge-based Development of Korea (Draft)", An Interim Report, June.

Murnane, Richard, and Frank Levy (1997),
"Teaching the New Basic Skills: Principles for Educating Children to Thrive in a Changing Economy", New York: Free Press.

OECD (1996),
Employment and Growth in the Knowledge-based Economy, Paris: OECD.

OECD (1998*a*),
Education at a Glance: OECD Indicators, Paris: OECD.

OECD (1998*b*),
Education Policy Analysis, Paris: OECD.

OECD (1998*c*),
Technology, Productivity and Job Creation: *Best Policy Practices*, Paris: OECD.

OECD (1998*d*),
Reviews of National Policies for Education: *Korea*, Paris: OECD.

OECD (1999),
"New Enterprise Work Practices and their Labour Market Implications", in OECD (1999), *Employment Outlook 1999*, Chapter 4, Paris: OECD.

OECD (2000),
Investing in Education: *Analysis of the 1999 World Education Indicators*, Paris: OECD.

Park, Se-Il (1999),
"Managing Educational Reform: Lessons from the Korean Experience: 1995-97", Washington, DC: The Brookings Institution.

Perris, Lyall (1998),
"Implementing Education Reforms in New Zealand: 1987-97", World Bank Human Development Network.

Reich, Robert, B (1991),
The Work of Nations, Chapter 4, New York: A.A. Knopf.

The Presidential Commission on Education Reform (Korea) (1997),
"Education Reform for the 21st Century", November, Seoul: The Presidential Commission on Education Reform.

Thomas, Christopher (1998),
"Back to Office Report and Aide Memoire: Korea Structural Reform in Higher Education", Washington, DC: World Bank, 17 August.

UNDP (1999),
"Human Development Report 1999", United Nations Development Program, New York: Oxford University Press.

US Division of Employment and Training (1998),
"Employment Projections for Industries and Occupations 1996-2006", Massachusetts.

Woo, Cheonsik, and Ju-Ho Lee (1999),
"Efficiency of Korean Education: Myth and Mission", Korea Development Institute, Working Paper No. 9906, April.

World Bank (1999*a*),
EdStats (Education Statistics Database), Washington, DC: World Bank.

World Bank (1999*b*),
"Corporate Survey of the Asian Financial Crisis", PREM, Washington, DC: World Bank.

World Bank (1999*c*),
World Development Indicators 1999, Washington, DC: World Bank.

Chapter 4

Ensuring a Dynamic Information Infrastructure

A. Background

The Korean Government and people have been quick to recognize the importance of the global networking revolution and to formulate a strategy to ensure that the country moves ahead to gain all possible benefits from that revolution. Korea has performed impressively in the information infrastructure sector in many areas – especially Internet access and mobile telecommunications. It has a well-educated workforce, a strong ICT manufacturing sector and a number of other advantages. However, to preserve this strong record, the country needs to move toward a more efficient model of ICT provision and services.

The 1999 White Paper *Cyber Korea* 21 notes that levels of information and the size of the knowledge gap greatly influence the productivity of individuals, businesses and economies. *Cyber Korea* 21 and related initiatives (including President Kim's New Year Policy Speech) lay out an ambitious set of targets to be met by 2002 to ensure that Korea becomes one of the top-ten nations in terms of information infrastructure and industry. *Cyber Korea* 21 highlights the importance attributed by the country's leaders to information infrastructure and sets out a number of policy goals. However, it suggests a role for the government that may be beyond what is appropriate in such a dynamic sector where technology and the speed and flexibility of the private sector may make strong government orchestration a hindrance rather than an asset.

This chapter provides a picture of recent growth in the Korean ICT sector – in both infrastructure and manufacturing – before turning to some of the problems facing the sector. It discusses the legal, regulatory and investment environment in Korea, arguing for a change of focus in regulation and support. It builds on the ideas laid out in *Cyber Korea* 21, but suggests a different focus in terms of regulatory and investment priorities. Broadly, it argues that Korea's high level of investment in information infrastructure has achieved impressive results. However, in common with the country's education sector, reform of the laws and regulations governing the IT sector could bring even higher returns to investment. Moving away from micro-management of research and investment with direct control of regulation toward a more hands-off model could lead to a more rapid roll-out of ICTs and ensure higher growth of the sector in the future.

B. Recent sector growth

1. Recent growth in telecommunications and the Internet

In 1980, telephone penetration in Korea was 7.3 lines per 100 people. By 1997, it was 44.4, almost reaching the OECD average of 48.9 telephones per 100 people (OECD, 2000). Between 1994 and 1998, the number of mobile phone subscribers grew from under 1 million to 14 million (see Figure 4.1), which, at 30.2 mobile phones per 100 people, was higher than the OECD average of 25. By the end of 1999, it is estimated that there were nearly 23 million mobile phones in Korea. Quality and cost of basic services is good, with the percentage of faults repaired within 24 hours the highest in the OECD area and rental charges well below the OECD average.[1] Today, one person in two owns a mobile phone.

Figure 4.1. **Past and predicted growth in mobile phones**

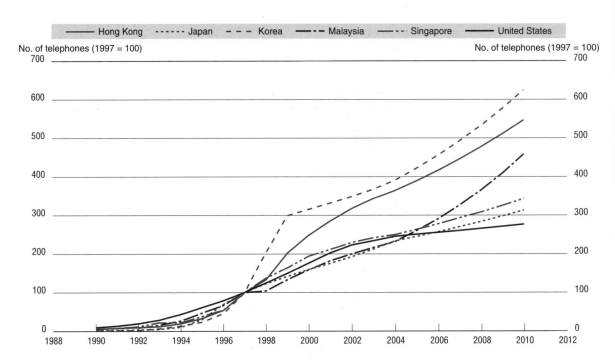

Source: Pyramid Research.

The number of Internet hosts is also increasing rapidly, with the country adding nearly 100 000 host sites between January 1999 and January 2000 (an increase of over 50%).[2] Estimates suggest there were already nearly 13 million Internet users in Korea in February 2000, up from just 3.1 million at the end of 1998 and only 1.6 million in 1997. At the end of 1999, 23.6% of Korea's population used the Internet – about the same proportion as in the United Kingdom and Chinese Taipei, and a significantly higher percentage than in countries such as Japan and Germany (both at around 15%). In 1999 alone, an estimated 2 million Internet-capable PCs were sold. The number of women Internet users in Korea is also growing (although not fast enough). Women users made up 16% of all Internet users in 1998, reaching 20% in 1999.[3] Korean advanced services are mostly supplied at low cost and with good public access:

- For consumers and small businesses using the Internet for less than 20 hours a month at off-peak times, a recent study by the OECD suggests that Korea's rates are the lowest among Member countries (OECD, 2000).[4]

- Public access is high. A recent survey of Web users worldwide found that Korea had one of the highest rates of access from school (at 29% of respondents) among the OECD countries (in Japan, the figure was 7%). Access from kiosks, libraries and cyber cafes accounted for over 20% of users – compared to less than 10% in Japan. This high rate of usage stems in part from the network of cyber cafes which offer service for less than USD 1 an hour.

- An information superhighway has been installed in 94 major areas of the nation, with a capacity of 2.5 gigabytes per second.

- In terms of use, a recent survey indicated that 56% of Korean users had purchased goods over the Internet from home (compared to a high of 72% in the United States, but only 37% in Spain).

- There are 400 domestic online shopping malls, with monthly average sales of some KRW 50 million to KRW 60 million (USD 41 700 to USD 50 000) as of January 1999.

According to *Cyber Korea* 21, employment in information and telecommunication companies climbed from 318 000 to 507 000 over the 1994-98 period, and public telecommunications revenue reached around USD 9.1 billion in 1997. Turnover in the ICT sector as a whole in 1999 is estimated to have increased by 20% to KRW 71 trillion.

2. A *strong private telecommunications manufacturing sector*

Korea is currently one of the world leaders in code division multiple access (CDMA) technology, with SK Telecom the world's largest CDMA operator and with numerous local vendors exporting CDMA equipment, especially handsets. In 1998, Korean companies exported USD 660 million of CDMA-based phone sets, and the nation's three biggest mobile telecom firms – Samsung Electronics, LGIC and HEI – planned to ship USD 1.62 billion of CDMA mobile phone equipment in 1999. Exports from the ICT sector constituted 67% of the nation's trade surplus in the first quarter of 1999 (*i.e.* USD 4.77 billion).

With nearly 40% of the global CDMA market, Samsung is expected to remain the world's largest supplier for the third consecutive year. The company has established a firm footing in the US market and recently concluded a USD 100 million contract with AirTouch Communications to provide its CDMA commercial system through 2003. In mid-July, Samsung won a three-year USD 160 million contract for its Internet phones from the Norwegian state-run telecommunications firm Telenor.

Arguably, government intervention enabled the benefits of scale with CDMA to be realized so quickly. The Ministry of Information and Communications (MIC) has focused on the development of an information infrastructure through local technologies, rather than relying on foreign vendors. Recognizing the potential of CDMA technology, the government pushed the industry to adopt that technology as the standard for mobile communications and to develop a local technology base. The lack of worldwide development of CDMA technology presented lower entry barriers for local companies to gain expertise with foreign companies through public and private research and development.

The Korean Government's interventionist approach to guiding the development of the industry has a parallel in Europe where governments adopted the GSM standard (although this was not part of an industrial policy, and merely aimed to ensure service integration). In contrast, wireless information infrastructure development in the United States has been hampered by multiple standards, resulting in an inability to interoperate among different carriers in the country as well as with the rest of the world. Nonetheless, despite the fact that it paid off, the choice to enforce a CDMA standard was a risky gamble. And it has involved costs – including the need to make significant payments to US patent owners. In Section E, we further discuss the limited applicability of this experience to other efforts at central planning at the micro level.

3. *Heavy investments in* ICT *infrastructure*

As Table 4.1 suggests, Korea invested heavily in its information infrastructure over the course of the 1990s. As a percentage of GDP, Korean investment was 0.8% in the first half of the decade and more than doubled to 1.85% of GDP in the second half. Compared with a number of other countries, this performance is very impressive. In percentage terms, in the second half of the 1990s, Korea's investment rate was nearly twice that of Hong Kong, and more than five times that of Japan.

Another distinguishing feature of Korean investment is that a high percentage of funding came from government. In many OECD countries, the public share in investment was zero in the late 1990s, while for Malaysia, Singapore and the United Kingdom, the public share was 5% or less. In Korea, 25% of investment came from the public sector (although this was down from 48% in the early 1990s). As a percentage of GDP, public investment in information infrastructure in Korea was higher than *total* investment in information infrastructure in Japan and the United Kingdom, and about similar to US total investment. As a percentage of GDP, public investment in the sector actually rose over the course of the 1990s.

Public investment in Korea has a fairly distinguished record compared to such investment elsewhere. Nonetheless, these figures raise the question of the returns to this investment. Does the perfor-

Table 4.1. **Korean investment in information infrastructure, 1991-99**

	Investment/GDP (%)		Public investment (% of total)	
	Average 1991-95	Average 1996-99	Average 1991-95	Average 1996-99
Hong Kong	0.58	0.98	0	0
Japan	0.14	0.34	0	0
Korea	0.80	1.85	48	25
Malaysia	1.12	1.04	6	5
Singapore	0.35	0.57	38	4
United Kingdom	0.23	0.35	2	2
United States	0.58	0.52	0	0

Source: Pyramid Research.

mance of the ICT sector suggest that this investment has collected the returns that might have been possible? While the country's performance has been notable, international comparisons suggest that the answer to this question is at best unclear.

C. Remaining challenges

While growth in service provision and the ICT sector has been striking in recent years, there is still some way to go. Perhaps most important in the future is the roll-out of more advanced equipment and services. The network digitalization rate in 1997 was 66.7% of fixed access lines, compared to an OECD average of 89.2%. In ISDN, as of May 1999, Korea Telecom had only 90 000 customers (although there were a total of 829 000 high-speed Internet subscribers in February 2000, about half of whom accessed using ADSL).[5] Continued advances in the quality of fixed service provision are central to further expansion of Internet use in the country, especially as consumers and businesses begin to demand an "always on" Internet environment (the ability to stay on line for extended periods).

Indeed, while we have witnessed impressive performance in the growth of the Internet over the last two years, this performance is less astounding from the perspective of a global revolution in Internet provision. Figure 4.2 shows that future growth in Internet use is predicted to be reasonably strong – but this is from a low base, with 0.03 Internet subscribers per capita in 1997, compared to 0.11 in Hong Kong, for example. And, although we have seen that advanced Internet services are developing, once again this is from a very low base. In August 1999, only two Korean banks had started offering Internet banking – and then only a limited range of services (although, again, there has been rapid progress over the last year).[6] We have seen that Korea has 400 (rather small) cyber malls – compare this to Japan's 7 000 and the United States' 450 000 (Ministry of Commerce, 1999).

Further, in common with the rest of the world, Korea has seen the emergence of a "digital divide" between rich and poor, more and less educated, men and women, rural and urban. For example, the usage rate among those with incomes above KRW 4 million is more than double that among people earning less than KRW 1 million. For farmers and fishermen, a usage rate of 7.3% compares to a usage rate of 64% among office workers.

Broadly, the picture is mixed for past returns to very significant levels of investment in the sector, and future estimates suggest that, in the absence of reform, returns to investment will remain lower than might have been the case. In mobile telephony, where competition prevails, returns appear to have been substantial. In other sectors with less competition, the evidence is less reassuring. How can better performance be ensured, especially in the provision of next-generation ICTs? Given the revolutionary nature of the sector, this report suggests that there is a vital role for government – especially in rolling out services to the disadvantaged – but that that role is different from that traditionally followed, involving more open competition and investment policies.

Figure 4.2. **Total Internet subscribers, past and predicted**

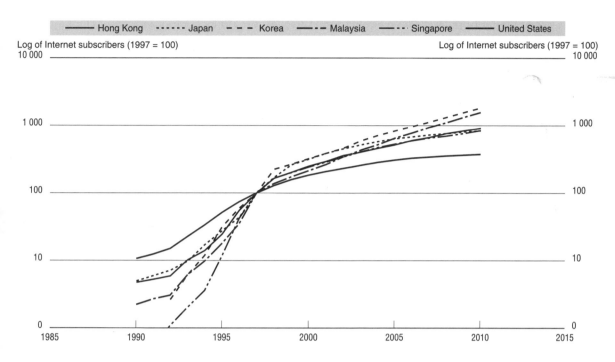

Source: Pyramid Research.

D. *Cyber Korea* 21: Targets and initiatives

Responding to both the need to increase returns to investment and to keep Korea at the cutting edge of information and technology roll-out, in 1999 the Korean Government launched *Cyber Korea* 21, together with a range of linked initiatives. Targets were laid out, including:

- Designating six strategic fields in which Korean firms will compete on the global stage: next-generation Internet, fiber-optic telecommunications, digital broadcasting, wireless communications, software and computers.

- Creating 1 million jobs and KRW 118 trillion of new production.

- Providing universal service access speeds of 2 Mbps.

- Nurturing more than 5 000 venture enterprises while doubling the proportion of Korean parts in IT equipment from today's 40%.

- Increasing the number of Internet users to 10 million by 2001.

In order to meet these ambitious targets, a range of government initiatives have been put forward:

- Reducing government ownership of Korea Telecom to 34% by the year 2000.

- Providing KRW 500 billion for the development of info-tech core technologies in a bid to end dependence on foreign technologies.

- Setting up a venture firm in information and communications and a venture co-operative to expand infrastructure for venture firms.

- Investing KRW 10.4 trillion in advanced information infrastructure and information technology by 2002.

- Removing impediments to e-commerce, including the Door-to-door Sales Act, and ensuring the validity and security of digital signatures.

- Connecting more than 10 000 schools to the Internet, providing computers free of charge to over 230 000 teachers and 200 000 classrooms, and providing free access to 50 000 children from poor families.

- Facilitating Internet PC purchases through a lump-sum payment at computer shops or through fixed installment savings accounts at the country's 2 800 post offices, with a target of 9 million new PCs, priced between USD 731 and USD 830.

- Building Internet Plazas in public facilities, including libraries and post offices, and providing kiosk access to certificate issuance services.

- Teaching 900 000 civil servants, 10 million students and 600 000 military personnel to use computers.

- Providing subsidized or free Internet training for groups including housewives, juvenile delinquents and the disabled.

- Providing e-mail accounts to all public servants and moving to a digital documentation and authorization system that will decrease work process times by up to 30% and digitize 80% of circulated documents.

- Digitizing the public procurement operations of 26 000 institutions (up from 556 in 1998).

The rest of this chapter explores the reform agenda in the light of *Cyber Korea* 21, and supports many of the recommendations of the report while suggesting a slightly different focus in terms of priorities and programs in specific areas.

E.   Next-generation policy and regulatory role for government

Korea has been moving toward a more liberal ICT sector. Excluding Korea Telecom, the last few years have seen the launching of initiatives to remove limitations on resale services, individual ownership and mergers and acquisitions, increase aggregate foreign ownership restrictions to 49% and allow the largest shareholder to be foreign. The liberalization of the local loop market in 1998 means that all parts of telecommunications services are now legally open to competition. There is an important role for oversight of dominant operators to ensure fair competition. However, as we will see below, other restrictions remain, both on the incumbent operator and on the freedom to compete with it. Further, many domestic players are concerned that the government's continued interventionist approach, which appears to favor large players, will reduce the creativity and vibrancy of the industry.

1.   *Opening up to domestic and foreign competition*[7]

Where more open and fair competition and less regulatory intervention has been allowed for significant periods, the results have been noteworthy – not least in the mobile sector and in third-party resellers for VoIP services. In 1995-98, mobile revenues increased nearly six-fold. With the removal of the government-imposed compulsory subscription period of two years for mobile in April 1999, the number of mobile customers shot up by 3 million in just one month. Low prices for Internet access are in part the result of fierce and relatively unencumbered competition between some 24 commercial and five non-profit ISPs, some of which offer free PCs in return for a three-year subscription.

Conversely, competition has been slow to deliver in fixed services. There have been only three new entrants in the fixed telecommunications infrastructure market (compared to 29 operators in Ireland). In the national long-distance market, Korea Telecom had a 91% market share in mid-1999, and held close to 100% of the local market. As a result, prices have remained high and service provision is weak. Between February 1996-99, the highest long-distance telephone charge in Korea dropped by 22%, compared to cuts of 50% in Japan or 46% in France. In addition, many advanced infrastructures (including ISDN) have been slow to roll out (although ADSL, for example, has grown more rapidly).

Independent regulation (discussed below) will be part of the solution to overcome these problems (by offering better support for competition). Another will be to lower the barriers to entry posed by very high license, annual and research fees levied on operators. At present, the MIC has the authority to recommend the level of investment required by each telecom provider to fund the research and development of telecommunications technology and other related projects. This is set by law at 3% of revenue (although the major operators did not meet this target in 1998 and it should be noted that the MIC plans to reduce these fees). A third method to overcome barriers to fixed competition would be to unbundle services. Presently, local loop network elements may not be unbundled, which means in effect that Korea Telecom's competitor in this market, Hanaro, must replicate all existing sunk plant at prohibitive cost. If unbundling remains prohibited, real competition will be unable to blossom in this market. A regulatory reform agenda to encourage stronger competition is laid out below.

2. Accessing international investment

Another important measure to ensure international competitiveness and improve domestic service provision is opening up access to international investment. As noted above, Korea has implemented WTO agreements on foreign ownership – and did so 18 months earlier than scheduled. However, remaining limits – including those on ownership of Korea Telecom itself – are a wasted opportunity to attract investment to a sector where there is clearly foreign interest. For example, British Telecom is expected to invest USD 500 million in LG, while Bell Canada International Inc. recently announced that it has agreed to invest USD 159 million to acquire about 23.6% of Hansol PCS Co. Ltd, one of three new Korean CDMA-based PCS operators.[8]

Thus, while Korea Telecom is extending its overseas reach (it recently acquired a stake in a Vladivostok telephone company), overseas investment in domestic ICT markets should also be encouraged. If the government is concerned about international control of the incumbent, other OECD countries have found the "golden share" approach to be more efficient than maintaining majority ownership.

Encouraging a greater role for private investment should allow for a reduction in government expenditure on service roll-out, which is also likely to increase efficiencies. Looking at the Internet backbone, government use of ICTs and universal access to advanced services, there remains a role for government support. However, this support should at the least be flexible and open to adjustment given changing circumstances. For example, capacity on the Korea Electric Power Corporation (KEPCO)'s dual-capacity fiber optic 2.5 gigabyte backbone was previously leased to 43 (out of 78) regional cable TV system operators purely for cable TV use. Recently, the backbone was spun off and new regulatory conditions will allow the cable companies to provide services including Internet over the network. Despite this "windfall" in backbone capacity, the MIC is yet to account for this change in its backbone investment plans.[9]

3. Reducing micro-regulation and control

The advantages of open competition suggest the gains that could be made by rolling back a range of micro-regulations in the sector. While government intervention to impose the CDMA standard suggests that early standardization is of benefit to consumers, counteracting this effect is the risk of "locking in" the wrong standard.[10] Therefore, any technology standards ought to be widely canvassed with the private sector before introduction and should remain as flexible and limited as possible.

Less defensible than standards from an economic standpoint are price regulations (with the exception of those designed to control monopoly rents). For example, the government's decision to forcibly abolish company subsidies on cellular phones seems an unnecessary intrusion into business practices. Cyber Korea 21 also suggests a continued government role in the pricing policies of Internet Service Providers, which appears unjustifiable in a very competitive market.[11]

Finally, it should be noted that, despite the need for a more market-friendly approach to regulation and restrictions, many areas will continue to need oversight and (possibly) a greater level of regulation. For example, entry into the specialized telecom service provider (STSP) market has been simplified, and this sub-sector is now very competitive. Recently, an STSP company went bankrupt, raising concerns about compensation of consumers – here, there is clearly a need for regulation. Also, there is currently a degree of cross-ownership between competing telecommunications providers – for example, DACOM owns 11% of Hanaro in the local telecommunications market. This should be discouraged, perhaps through regulation.

4. Independent regulation

Perhaps the most important change that might be made to the regulatory environment is the creation of an independent regulatory agency. Korea is one of only two countries in the OECD which does not have such a body. The Korean Communications Commission (KCC) is charged with ensuring fair competition in the sector through arbitration of disputes, fact finding on unfair practices, examination of competition-related rules and regulations, and proposing corrective measures against unfair practices. This body is nominally neutral and can, under limited circumstances, over-rule the MIC. For example, when it arbitrates between companies, its decisions cannot be overturned by the Minister. However, the KCC lacks independent legal status (coming under the supervision of the MIC) and has but 20 employees who remain part of the MIC's management structure. Its powers are largely indirect and advisory; its role reactive rather than proactive.

The problems created by this lack of independence are illustrated by the history of STSP competition. Since the government liberalized the international call market, STSPs have mushroomed and their share of the market is now estimated at 15 to 20%. In the wake of broader competition, international calls now cost one-half or one-third of previous prices. However, the STSPs claim that KT is demanding that newcomers pay three to five times more for using KT equipment than do existing carriers such as Dacom and Onse Telecom, while the nation's public network operators, who once enjoyed an oligopoly in the international call market, complain in turn about the new competitors. They charge that the STSPs are getting a free ride, demanding to use facilities in which backbone network operators have made huge investments. Without an independent regulator, at the very least it will be hard to convince all parties that rules are being enforced in a fair way.[12]

Only with an independent regulator are all sides likely to feel equally treated – a vital step toward encouraging investment and competition in the sector. An independent KCC, separate from the MIC and with its own staff and budget, would more effectively achieve market confidence and transparency. The KCC would need powers of licensing, price control and interconnection, overseeing policies on universal service and the implementation of other regulatory safeguards.

5. A telecommunications regulatory reform agenda

A recent OECD report (OECD, 2000) laid out a broad agenda for regulatory reform in the telecommunications sector. Beyond independent regulation and fewer price controls in competitive markets, it suggests that priorities for improved independent regulation in this area should include:

- Shifting to the long-run average incremental cost methodology (away from the fully distributed cost methodology) in calculating interconnection costs. This would help avoid manipulation of distribution of common costs and penalize inefficiencies in the former monopoly carrier. Currently, while KT interconnection charges are in the mid-range of OECD countries, they are high relative to retail prices and revenues.

- Creating a universal service costing methodology and moving toward a technologically neutral universal service fund as the vehicle for supporting roll-out.

- Simplifying the licensing regime by removing existing distinctions between fixed provision of local and long-distance services and perhaps moving toward a class-licensing system.

- Simplifying the registration process for licenses by: moving to *ex post* examination of licensee standards and agreements rather than *a priori* examination wherever possible; allowing for rolling applications rather than offering a twice-yearly application window; reducing data requirements for licensees; and removing R&D requirements on licensees.

- Rolling out IMTS-2000 mobile licenses (the MIC is considering an auction process as one of the options in selecting service providers). Pre-qualification for this process might enhance competitive conditions in the market.

- Adopting price caps for dominant local loop, leased line and long-distance provision (this is already being reviewed by the MIC), and removing price restrictions in the competitive mobile market.

- Implementing arrangements for unbundling.

- Introducing number portability to encourage competition by allowing users to change service providers without changing numbers (pledged by the MIC for this year).

- Reducing government involvement in accounting rate negotiations with foreign operators based in a competitive market. Currently, the MIC approves a single accounting rate for all carriers operating between Korea and a second country. This is one reason for very high accounting rates (leading to expensive international calls) in Korea. The US accounting rate with Korea is 71 cents, compared to the US-France rate of 20 cents or the US-Japan rate of 28 cents.

- Minimizing overlap in the roles of the KCC and the Korea Free Trade Commission which enforces the Fair Trade Act on all sectors including telecommunications (for which there are no exemptions).

- Increasing the transparency of the policy formulation process. Progress has already been made here, with the government holding public hearings on some regulatory proposals and providing notice of changes to the Telecommunications Business Act. Major telecommunications legislation is now reviewed by the Information and Communication Policy Deliberation Committee made up of outside experts. Nonetheless, user groups, including the Federation of Korean Industries, could be encouraged to play a more active role in the policy formulation process.

- Moving toward joint independent operation with broadcasting regulators (the MIC has suggested that such an approach be implemented over the next year).

6. *Legislation and support for research and the manufacturing sector*

We have seen that the government mandates research expenditures for telecommunications services suppliers. *Cyber Korea* 21 plans to step up government support for investment in technology and manpower to small venture firms in the information and telecommunications industry.

There are particular concerns about the MIC having the joint roles of consumer protection and industry promotion – most regulatory bodies in the OECD are solely concerned with consumer interests. This problem would be reduced with the granting of greater independence to the KCC. However, it should be noted that promoting the domestic telecommunications manufacturing industry does not necessarily have a positive effect on the service quality of domestic telecommunications. Indeed, if domestic promotion involves discriminating against superior imports, the opposite effect might be achieved. For example, Cisco Systems has already captured 80% of the domestic Internet equipment market. It might be a mistake to encourage a domestic competitor under these circumstances.

Instead, the manufacturing sector ought to be opened up, creating a framework in which manufacturing and services can develop in a competitive environment. This is especially true in the case of trade. Korean telecommunications manufacturers have greatly benefited from healthy export markets and the capability to exploit them. At the same time, however, European mobile telecommunications conglomerates have begun building CDMA handsets. In response, Korea should liberalize its own telecommunications manufacturing industry, both to ensure that manufacturers remain internationally competitive at least on the easier home market and to put itself in a position to argue for the elimination of trade barriers in telecom fields elsewhere in Asia, Latin America and Africa.

Where a role remains for support through targeted investment, the Scottish model of venture capital support (rather than direct intervention in management decisions) might provide a good model of a relatively hands-off approach to industry promotion. Further, foreign investors are showing great interest in Korean manufacturing operations. The recent investment by Philips in LG-LCD, one of the world's largest liquid crystal display manufacturers, came about partly because of the high level of manufacturing technology of the Korean counterpart. Korean companies have abundant skilled labor, and development costs could be much cheaper in Korea because the wages of Korean engineers are lower than those of other OECD countries. Where it is available, tapping foreign investment and expertise is likely to be a more efficient mechanism for promoting a strong ICT service sector than widespread government investment and closing off the manufacturing sector from foreign competition.

7. Legislation and regulation for electronic commerce

There is also the need for wider reform to prepare businesses for the digital age. Venture capital has an important role to play in supporting the entrepreneurs who typify Web development – small and relatively inexperienced in business and financial management. Governments in other countries (such as Scotland), as mentioned earlier, have supported venture capital funds to develop the ICT industry, similar in some respects to the funds to be set up in Korea. More important to further development of ICT industries, however, is the broader legislative and regulatory environment. Full competition in telecommunications and Internet service provision is perhaps the first step. Beyond that, the Ministry of Finance and Economy's decision to streamline laws that negatively impact upon SME operations is a significant move, as is the government's overhaul of 106 laws and regulations that will affect the growth of the knowledge-based economy. More narrowly, a number of considerations relating particularly to e-commerce are already being addressed through Korea's Basic Act on Electronic Commerce, the Electronic Signature Act and the Act on Telecommunication Infrastructure Protection. A number of steps remain to be taken, however.

As we have noted, Korea's electronic commerce is expected to grow from KRW 150 billion (USD 125 million) in 2000 to KRW 3.7 trillion (USD 3 billion) in 2002, so the government needs to prepare for rapidly expanding Internet transactions. Tax incentives for online transactions are not necessary, but a different approach to taxation might well be required – for example, modifying the VAT system. It is also important to study ways of protecting consumers and ensuring the privacy of information, with a need for electronic signatures and certification systems – especially at public terminals.

National initiatives from around the world suggest a range of activities for government in this sector (UK Government, 1999; Computer Systems Policy Project, 1999; Ireland Advisory Committee on Telecommunications, 1998; OECD, 1998):

- *Encouraging the foundation of a digital economy*, including: the mapping of emerging clusters of e-commerce and examining whether their development can be facilitated; encouraging training, supporting industry mentoring and partnering – especially for and with SMEs, as with Korea's Electronic Commerce Resource Centers; ensuring the equivalence between digital and written documents; supporting electronic signatures, payment systems and certification systems (a measure supported in *Cyber Korea* 21); and ensuring widespread broadband access.

- *Enhancing trust*, including: the provision of a secure public key infrastructure (an adaptable and secure form of encryption), especially for government-private sector and government-citizen relations; allowing users free choice of cryptographic methods (and using them to protect critical infrastructure); guaranteeing the rapid and widespread prosecution of electronic fraud and other electronic crimes; supporting "parents' Web sites" and content-filtering software; and ensuring that government itself adopts "best practice" information and privacy policies.

- *Ensuring sustainability*, including: adapting tax laws to the new commercial environment; promoting widespread access both within and outside government (a key part of the *Cyber Korea* 21 initiative); and preserving the value of intellectual property through enforcement of the 1996 World Intellectual Property Organization treaties.

While the passage of the Electronic Signature Law in July 1999 was a step in the right direction, a number of barriers to e-commerce remain. For example, under current laws, only the Korean Fund Transfer Center and the Korean Fund Transfer Company are allowed to execute foreign exchange business, inhibiting a number of key Internet financial services. Similarly Internet brokerage operations are not allowed outside the insurance business. The Door-to-door Sales Act (DDSA), which requires all firms conducting business over the Internet to have a business presence in Korea and to comply with Korean law, also acts as a bottleneck. It is a commendable part of the plans laid out in *Cyber Korea* 21 to repeal this law. In general, a broad survey of laws governing commerce should be conducted, and those laws that would unintentionally or unnecessarily restrict e-commerce should be amended so that Internet firms can do business freely. Foreign standards on e-commerce transactions should be mutually recognized, while ensuring the protection of the Korean consumer.

The issue of taxation of e-commerce is a particularly complex one which is still being worked out worldwide. Now is the time, however, to move toward novel taxation systems that account for the growth of e-commerce – expected to equal 7.2% of GDP in the United States by 2001, for example. A second consideration is international harmonization of rules regarding the taxation of e-commerce, although this issue is slightly less urgent as the WTO predicted in 1998 that only USD 60 million in international trade would be conducted over the Internet in 2001 (Boyle *et al.*, 1999). Korean tax administrators face a number of challenges in an environment where it is difficult to establish the place of residence of both the seller and buyer (or consumer) of a product:

- Preventing tax avoidance (for example, with mail-order goods) and double taxation (on some services).
- Discouraging new forms of tax fraud and evasion – both intentional and accidental. This would involve enhanced capabilities in tax and police departments as well as information campaigns for (in particular) SMEs.
- Ensuring consistent treatment of cross-border transactions.
- Ensuring privacy.
- Limiting transaction costs – this could be achieved in part through the use of advanced software and database technologies.

Some solutions being examined worldwide include licensing suppliers to support (supplier) residence-based taxation. This, in turn, might be streamlined through automated transaction reporting systems embedded in software – perhaps browsers – and tax forms redesigned to obtain data on electronic trading. These solutions raise a number of privacy and economic efficiency concerns, however. The OECD has convened a number of discussions on e-commerce issues and has concluded that VAT should be taxed according to the point of consumption. These discussions are ongoing.

One further important element will be to move toward a common global standard of classification of services similar to that for goods maintained by the World Customs Organization (the Harmonized Commodity Description and Coding System). The United Nations Central Products Classification Scheme might be a system on which to build (the United States Census Bureau appears ready to use a classification system based on UNCPC, for example) (US National Tax Association, 1999). It will also be important over the medium term for Korea to take a significant role in international discussions on the direct tax treatment of international e-commerce transactions, especially relating to VAT and customs duties on small consignments of imports.

F. Ensuring access and overcoming the digital divide

While much of this chapter has discussed the need for a lighter government hand – less regulation, greater distance between bureaucrat and entrepreneur – there are a number of vital roles that the government retains in the information economy. Some of these roles will require an even greater level of activism. Perhaps most importantly, in order to ensure that the digital revolution benefits all rather than being reserved for an elite, it is crucial to ensure access to the new information technologies – both for the individual and for small companies wishing to bid for government contracts on line, for example.

A first step toward universal access to information infrastructure is the universal service obligation for basic telephone services – although Korea, with telephone penetration rates above the OECD average, should find this challenge easy to address. At least in the short term, it is likely that the incumbent could shoulder the responsibility at limited cost. In the longer term, as mentioned above, its universal service compensation fund should be competitively and technologically neutral. However, beyond basic services, universal access to more advanced equipment is likely to be better funded directly through general government revenues rather than by imposing requirements on telecommunications operators.

Improving equity of access to the Internet will involve the widespread training and access policies outlined in the *Cyber Korea* 21 White Paper – initiatives such as connecting 10 400 schools; teaching civil servants, housewives, students and military personnel to use computers; building Internet Plazas; and facilitating Internet PC purchases.

There are a number of models around the world of training programs designed to ensure ICT technological literacy. In Singapore, there are computerized community clubs, staffed by volunteer trainers, in the proximity of every public housing complex (where more than 80% of the population lives). The city-state has also encouraged private companies to sponsor mass-training exercises. It is important that such exercises reach out and are flexible to the needs of the disadvantaged – the poor, the elderly and the disabled.

Technological literacy is only the first step, however, and there is a large role for government to support more advanced training. Ireland has introduced training programs for future Net entrepreneurs in the business skills that they will need to run their own Internet start-up. Finally, although English is losing its position as the monopoly language of the Internet, it is likely to remain dominant. While there might be a role for government to support Korean-language content in some sectors (for example, public health), the networked economy also increases the importance of foreign-language skills.

Training is clearly only useful if people have access to networked computers. A model for the provision of access, used in this case for the provision of public telephones but adaptable to public Internet provision, is the Chilean subsidy auction system. In 1994, Chile set up a limited-life fund to support the provision of the first payphones to remote and rural areas. Companies were asked to bid for the lowest subsidy that they would accept to provide service. Within two years, the fund had achieved 90% of its roll-out objectives, using only about half of its USD 4.3 million budget – largely because it received bids to provide service with no subsidy to about half of the chosen locations. Just over USD 2 million in public funds had leveraged USD 40 million in private investment to install telephones in 1 000 localities at about 10% of the costs of direct public provision (Wellenius, 1997). Such methods, by supporting public provision through tapping private entrepreneurialism, are likely to achieve more than the direct government provision suggested in the *Cyber Korea* 21 report – and at lesser cost.

G. Using ICTs to improve the performance of government

Another unique role for government in the information economy is to introduce ICTs into the 21% of GDP that it is directly responsible for. *Cyber Korea* 21 lays out some impressive targets for using ICTs to improve the performance of government – for example, providing e-mail accounts to all public servants and digitizing public procurement operations and document circulation.

The range of ICT applications in government suggested by international experience is huge:

- In Ireland, the Department of Social Welfare supported decentralization efforts through more than a dozen computer applications that improved information transfer and co-ordination.
- In the United States, "collusion detection software" is being used to root out impropriety in public bids.
- In Spain, smart cards are used to collect unemployment benefits at kiosks and to check on job vacancies and job opportunities.
- In Malaysia, government departments have collaborated on a networked system to facilitate land-resource management (Heeks, 1998).
- Thirty million people a month access New York City's Web site where they will soon be able to pay local taxes, municipal bills and parking tickets, as well as secure permits, look for govern-

ment jobs, bid on government auctions and obtain government and community information and news.[13]

- The Estonian Association of Telecottages is preparing to exploit IT to increase governance, planning a pilot project of virtual municipalities based on *Kuusamo* in Finland. There, villagers are able to access information about local government budgets, contacts, laws, transport, health and counseling services. They will also be able to access forms and documents.

International experience also suggests some important lessons, however. First, ICTs can only play a part in reform of government processes. While they allow for greatly increased information processing and flow, only with broader institutional reform will the full benefits of ICT introduction be felt. Second, it is important to walk before running. Bringing ICTs into any workplace is a complex, time- and resource-intensive activity. Moving from no computerization to full government on line in a single stage is a difficult proposition. Third, and perhaps most significant, it is important not to leave clients behind. Online government can only be equitable if all citizens are on line. If ICT access does not reach the vulnerable – groups such as the old, the poor or the disabled – the very people who need government services the most are likely to be denied them. This only emphasizes the importance of overcoming the digital divide.

Seattle might provide a model here. The city is bolstering its public access network with remote wireless Internet connections at eight storefront "little city halls" and installing Internet gateways for 85 PCs at five community centers throughout the city. Officials have also hired a consultant to draw a technology map that will tell residents where they can find public Web connections from locations as diverse as senior centers and neighborhood convenience stores.

Finally, there is need for greater co-ordination within government. The proliferation of government initiatives on the knowledge economy and society within individual government departments in Korea will further slow implementation of *Cyber Korea* 21 unless increased co-ordination can be ensured (see Chapter 7). In the United States, co-ordination between the White House, the Department of State, the Department of Commerce and the FCC, among others, has been a vital element in fostering the benefits of the Internet and electronic commerce for businesses and consumers.

H. Conclusions

Korea has made impressive strides in rolling out and updating its information infrastructure. This has involved not only the infrastructure itself, but also the manufacturing sector that creates it. However, in such a fast-moving field, strides are not enough – worldwide, the sector is changing and expanding by leaps and bounds. A number of *Cyber Korea* 21's goals – creating new jobs and production, increasing the bandwidth of universal service, nurturing venture enterprises, increasing the number of Internet users – are admirable. However, the sector's progress is constrained by an unwieldy and outdated regulatory and legal structure which will have to be streamlined. Without reform, the goals of *Cyber Korea* 21 will be unachievable – or will only be achieved at great cost. Moving toward independently regulated competition will allow for the more efficient upgrading of technology and provision of services. At the same time, to ensure that all parts of Korean society move forward together, the country's plans for inclusion and use of ICTs in governance must be given the highest priority. Revisiting *Cyber Korea* 21 and related targets, what does this mean for the country?

- Reducing government ownership of Korea Telecom is an important step – but one that could be taken further. It is only the first of many needed steps to reduce the role of government in this sector. There is a need to revisit regulations on foreign ownership and licensing, to reduce direction of research and pricing, to push forward on unbundling and a range of other issues.

- Government has a role to play in investing in advanced information infrastructure – although major support should be limited to the backing of private provision of services to government agencies and departments. These plans should be highly flexible, taking into account, for example, the "windfall" of the KEPCO network.

- Providing support to end dependence on foreign technologies, setting up a venture firm in information and communications and a co-operative to expand infrastructure for venture firms are all complex undertakings. In particular, the goal of "ending dependence on foreign technologies" is likely to be overly ambitious. It is unlikely that Korea will have a strategic comparative advantage across the entirety of the ICT sector. While there might be a small government role in encouraging the growth of SMEs in the Internet sector, a hands-off approach is likely to produce better results than detailed targeting of firms and technologies. The government's primary role in creating a dynamic information infrastructure is to ensure that the market conditions and incentives are in place to allow for private sector competition and investment.

- Removing impediments to e-commerce and ensuring the validity and security of digital signatures are important first steps. The Korean Government will need to examine a range of other issues – such as taxation of e-commerce and the provision of public key infrastructure – in order to enhance its ability to move transactions on line.

- Connecting schools, providing computers free of charge to teachers, facilitating Internet PC purchases and building Internet Plazas in public facilities are valuable steps toward the goal of expanding access. For the roll-out of ICT infrastructure, a market-conforming measure such as the Chilean auction scheme might provide a model mechanism for efficient provision, and might be a more responsive and targeted mechanism than blanket subsidies.

- Teaching civil servants, housewives, students and military personnel to use computers are also laudable goals. Use of the Singapore model might further expand access to training.

- Providing e-mail accounts to all public servants and moving toward a digital documentation and authorization system, along with digitizing public procurement operations, should greatly increase the efficiency of government. However, it becomes all the more important to ensure that the disadvantaged are not excluded from services or SMEs from bidding for contracts – again, this emphasizes the need to ensure equality of access and training.

With the help of such reforms, Korea can put itself in a leading-edge position in terms of public and government access and use of ICTs.

Notes

1. Korea did not do as well in increasing the number of personal computers per 1 000 persons between 1994 and 1998. By 1998, the number was 157 PCs per 1 000 persons, compared to 458 in Singapore and an OECD average of 256.

2. In 1998, Korea had 38 Internet hosts per 10 000 people, compared to 188 in Singapore and an OECD average of 254.

3. This figure is clearly too low and points to one of the costs arising from discrimination mentioned elsewhere in the report. The recently announced initiative on subsidizing training for home-makers should help to solve this problem.

4. Although in the case of a "basket" encompassing PSTN connection and usage fees plus ISP charges for users staying on line for 40 hours or more a month, Canada and the United States are still the cheapest markets at peak times.

5. OECD (2000); and Ministry of Information and Communication, Paris.

6. HSBC (1999). Local banks are adopting the Internet banking system, however, turning it into one of their main business objectives this year, by offering diverse services including loan applications. Kookmin Bank has seen the number of its Web-banking customers top 100 000 and 1 000 new clients are registering every day. Shinhan Bank receives 1 000 online loan applications daily, as opposed to just 400 at bank counters.

7. This and the following four sections draw heavily on OECD (2000).

8. The global rush to invest in Korean Internet companies has increased since last spring, when Web giants began to forge strategic ties with Korean companies. Amazon.com, one of the largest US online music and book retailers, formed a partnership with Samsung Corp., which recently announced that it would transform itself into an Internet-oriented firm through a strategic alliance with America Online. In March, Microsoft launched a Korean version of msn.com, its Web search and content aggregation service, while Lycos Inc. will open a Korean-language portal in July through a joint venture with Mirae Corp., investing USD 3 million for Net services. A US online brokerage, E*Trade, plans to team up with LG Securities to offer a joint stock trading service in Korea. These foreign investors will bring needed talent and expertise to Korea's nascent advanced Internet services.

9. Although it should be noted that in the wake of the Asian crisis, the MIC has already scaled back its planned investment from KRW 63 trillion to KRW 10.4 trillion by 2002.

10. A commonly cited example of sub-optimal "locking in" is the VHS standard of video systems, visually inferior to the Betamax system but which more quickly achieved a critical mass of users. Although this example was a private sector failure, government enforcement of standards can increase the risk, especially in a rapidly developing sector such as telecommunications.

11. Increased competition is likely to produce losers as well as winners – and might even increase the rate of bankruptcies in the sector. If these are the result of fair competition, however, they are an important part of the market process.

12. Other problems include the fact that Korea Telecom frequently competes against private operators such as SK Telecom or Samsung Electronics for research funds, licenses or services – and there are complaints of unfair treatment. Another example is the selection process of the broadband wireless local loop spectrum. Newly selected corporations were charged a fee for the right to bid, while Korea Telecom and Hanaro Telecom were selected with no charge six months earlier. Korea Telecom and Hanaro were thus given preferential access to a valuable spectrum resource at no cost.

13. See the Web site at *http://www.ci.nyc.ny.us/html/dot/html/ssss/home.html*.

References

Boyle, Michael, P. Peterson, John M. Sample, William J. Schottenstein, Tamara L. Sprague (1999),
"The Emerging International Tax Environment for Electronic Commerce", *Tax Management International Journal*, Vol. 28, pp. 357-382, 11 June.

Computer Systems Policy Project (1999),
Advancing Global Electronic Commerce: Technology Solutions to Public Policy Challenges, Washington, DC.

Heeks, Richard (1998),
"Information Age Reform of the Public Sector: The Potential and the Problems of IT for India Information Systems for Public Sector Management", Working Paper Series No. 6, IDPM, University of Manchester.

HSBC (1999),
Securities Sector Report: Korean Internet Plays, London: HSBC, August.

Ireland Advisory Committee on Telecommunications (1998),
Report to the Minister for Public Enterprise, Dublin: Government of Ireland.

Ministry of Commerce (Korea) (1999),
"Maximizing the Benefits of Electronic Commerce: Korea's Initiatives", Seoul.

OECD (1998),
"The Role of Telecommunications and Information Infrastructure in Advancing Electronic Commerce", internal working paper, Paris: OECD.

OECD (2000),
Regulatory Reform: Korea, Paris: OECD.

Pyramid Research Inc. (1999),
Information Infrastructure Indicators, 1990-2010, prepared for infoDev, Washington, DC: World Bank, available online at *http://www.infodev.org/projects/375/fin375.htm*.

UK Government (1999),
e-commerce@its.best.uk: A Performance and Innovation Unit Report, London: HMG.

US National Tax Association (1997),
Communications and Electronic Commerce Tax Project, Final Report, Washington, DC.

Wellenius, Bjorn (1997),
"Extending Telecommunications Service to Rural Areas – The Chilean Experience", World Bank Viewpoint Note No. 105, Washington, DC: World Bank.

Chapter 5

Improving the Korean Innovation System

A. Introduction

Korea's innovation system (KIS) remains largely based on the catch-up model. In the past, this model was quite effective in achieving Korea's development goals in a short period of time and enabled the country to rapidly master a wide range of industrial and technological activities. However, this model has its limitations. Private sector R&D, the lion's share of gross R&D spending, remains mainly oriented toward short-term technological development, while the majority of public sector R&D programs are highly mission-oriented but weak in terms of diffusion. Overall, Korea's development strategy in the S&T area has strongly favored the rapid exploitation of mature and advanced technologies for market expansion in selected sectors, over a broader strengthening of its knowledge base. The KIS has had relatively weak global linkages, except through arm's length ways of acquiring foreign knowledge such as technology embedded in equipment goods and components and formal technology licensing contracts. Until recently, Korea has made little effective use of technology acquisition through foreign direct investment (FDI), strategic alliances or joint research programs.

Today, Korea has come up against the limits of this strategy. In a world where trade, investment and production are increasingly globalized, the capacity to develop, acquire, diffuse and commercialize knowledge is becoming the major source of competitiveness and growth. In the knowledge-based economy, production and innovation increasingly require complementarity between "partners" at national and international levels. The holistic concept of the national innovation system (NIS) highlights this requirement.[1]

In many respects, Korea's potential research capability compares well with that of other OECD countries. High R&D spending and relatively well educated human resources are among its assets. However, the KIS is weak in terms of systemic linkages and interfaces among innovation actors. Korea's transition to the knowledge-based economy will therefore require a reorientation of its innovation system. A number of questions need to be examined. Is the current research orientation appropriate for supplying the KBE with the necessary knowledge base? Is its current configuration adequate for making efficient use of available resources? Is its current incentive structure suitable to effectively respond to changes in technology, market and social conditions? In the wake of the recent financial turbulence, the country has recognized the importance and urgency of this issue, and has introduced several important policy measures. While some are very positive, others need to be more carefully designed from a systemic perspective.

B. Profile of science, technology and innovation activities

1. R&D *expenditure and output*

Korea ranks very high among OECD countries in terms of R&D intensity (R&D as a share of GDP), along with Finland, Japan, Sweden, Switzerland and the United States. It has the lowest government share of R&D financing, followed by Japan, pointing to the prominent role of the private sector in R&D financing and performance. R&D intensity in the business sector is very high, with only Finland and Sweden showing higher intensities. In contrast, the share of R&D performed by the Korean higher education sector is very low and, in terms of researchers per 10 000 labor force, the country ranks close to the EU average.

Korean R&D spending is high and in 1998, its total R&D budget in current PPP USD ranked seventh in the OECD, roughly equivalent to that of the Nordic countries (Denmark, Finland, Norway and Sweden) taken together. However, Korea's R&D activities can be characterized as "high input" with a "biased composition of output"; the generation of codified knowledge (*e.g.* in the form of patents and publications) is relatively low compared to that of knowledge embodied in traded goods (Table 5.1). The number of scientific and technical articles per unit of GDP is quite low. Other patent-related "output" indicators shown in Table 5.1, such as technological strength and technological intensity, confirm that Korean R&D efforts contribute less to the generation of codified knowledge.

This can be seen from the pattern of Korean patenting in the United States. Patenting has risen, indicating the increased innovative capability of Korea, and Korean patents represented 2% of all utility patents granted. However, of all patents of foreign origin, Korea ranked fifth, with 5% of the total. Patenting is mainly concentrated in the *chaebol*, with Samsung playing the dominant role. Samsung Electronics

Table 5.1. **Income and technological performance, 1998[1]**

	Income level, 1996 GDP per capita as % of OECD average	Indicators of scientific and technological activities							
		Gross domestic expenditure on R&D as a % of GDP, 1998[1]	Researchers per 10 000 labor force, 1997[1]	Government-financed R&D as a % of GDP, 1998[1]	Government financing of R&D as a % of total R&D, 1998[1]	Business expenditure on R&D as a % of business GDP, 1998[1]	Scientific and technical articles per unit of GDP[2]	Technological strength per USD of R&D[3]	Technological intensity[4]
United States	140	2.8	74	0.8	30.6	2.3	20	410	10.4
Norway	128	1.7	76	0.7	42.9	1.3	21	. .	. .
Switzerland	126	2.7	55	0.7	26.9	2.3	37	. .	. .
Japan	121	2.9	92	0.5	18.1	2.3	15	354	10.6
Iceland	118	2.0	91	1.1	55.9	1.2	23		. .
Denmark	117	1.9	61	0.7	35.7	1.9	31	87	1.6
Canada	114	1.6	54	0.5	31.9	1.3	25	203	3.3
Belgium	112	1.6	53	0.4	26.4	1.4	20	111	1.8
Austria	111	1.6	34	0.7	44.6	1.1	18	125	1.9
Australia	107	1.7	67	0.8	46.0	0.8	24		. .
Germany	107	2.3	59	0.8	35.6	2.0	21	215	5
Netherlands	106	2.1	50	0.8	39.1	1.4	31	170	3.5
France	103	2.2	60	0.9	40.2	1.8	20	115	2.7
Italy	102	1.0	32	0.5	51.1	0.7	13	101	1
Sweden	100	3.9	86	1.0	25.2	4.4	41	147	5.3
United Kingdom	98	1.9	51	0.6	30.8	1.6	29	160	3.2
Finland	96	2.9	83	0.9	30.9	2.9	35	114	2.7
Ireland	92	1.4	51	0.3	22.2	1.3	16	69	1
New Zealand	88	1.1	45	0.6	52.3	0.4	29	. .	. .
Spain	77	0.9	33	0.4	43.6	0.5	16	21	0.2
Korea	**72**	**2.9**	**48**	**0.7**	**22.9**	**2.5**	**5**	**25**	**0.7**
Portugal	70	0.7	27	0.4	68.2	0.2	7	8	0
Greece	67	0.5	20	0.2	46.9	0.2	16	. .	. .
Czech Republic	64	1.3	24	0.5	36.8	1.0	15	. .	. .
Hungary	47	0.7	28	0.4	56.2	0.3	20	115	0.7
Mexico	36	0.3	6	0.2	71.1	0.1	2	15	0
Poland	35	0.8	32	0.5	59.0	0.4	17	. .	. .
Turkey	30	0.5	8	0.3	53.7	0.2	4	. .	. .

1. Or latest available year.
2. Scientific and technological articles per billion USD of GDP. See National Science Foundation (1998).
3. Technological strength is determined by multiplying the number of patents by an index of their impact. This index measures how frequently a country's recent patents are cited by all of a current year's patents. The patents refer to those granted at the US Patent and Trademark Office. Data are from CHI Research.
4. Technology intensity compares the technological strength of a country with its GDP expressed in PPP USD. See OECD (1998a) for details.
Source: OECD calculations, based on the Main Science and Technology Indicators database; CHI Research; National Science Foundation (1998); and OECD (1998a).

Table 5.2. **Flows of R&D funds in Korea, 1997**

KRW 100 million

Financed by	Performed by:							
	National and public research institutes	GRIs	Other non-profit orgs.	National and public universities	Private universities	Government-invested organisations	Private enterprises	Total
Government	**99.67**	**82.87**	19.85	**64.47**	30.86	1.21	4.67	25 897.88 (21.25)
Government-funded institutes	0.08	4.47	0.90	6.58	5.24	0.12	0.47	1 808.43 (1.48)
Other non-profit organisations	0.01	0.13	**40.87**	2.57	1.57	0.63	0.36	1 326.81 (1.09)
National and public universities	0.00	0.01	0.07	8.42	0.13	0.00	0.02	393.73 (0.32)
Private universities	0.00	0.01	0.03	0.18	**47.99**	0.00	0.00	4 021.29 (3.30)
Government-invested organisations	0.05	8.38	31.37	3.25	1.31	**97.90**	0.11	10 355.34 (8.50)
Private enterprises	0.17	4.13	6.90	14.45	12.14	0.09	**94.31**	77 936.62 (63.96)
Foreign sources	0.01	0.01	0.01	0.09	0.76	0.05	0.06	117.96 (0.10)
Total	3 805.4 (3.12)	15 106.2 (12.40)	1 777.3 (1.46)	4 361.5 (3.58)	8 354.4 (6.86)	8 363.4 (6.86)	80 089.5 (65.72)	121 858.1 (100.00)

Note: Numbers (except total) are percentage shares for each column. Numbers in parenthesis under totals are percentage shares in total.
Source: MOST (1998).

Co. Ltd. patented the most (40%), followed by Daewoo electronics (9.7%), LG Semiconductor Co. Ltd (7.2%), LG Electronics Inc. (6.6%), and Hyundai Electronics Industries Co. Ltd. (6.5%). Patenting by Samsung almost exclusively concerns electronics, with little breadth to the patenting, either by company or by sector (US Patent and Trademark Office, 1998).

2. *Flows of R&D funds*

Flows of R&D funds in Korea show some peculiarities. First, shares of self-financing are very high in both the public and private sectors. Government is the main financier of government research institutes (GRIs), at more than 80%. Industry also shows a very high share of self-financing, at more than 90%. The higher education sector shows the most diversified sources of funding, although this is mainly the result of weak public support for university research.

Table 5.2 shows that private enterprises account for the lion's share of both R&D investment and spending; the government sector's share is moderate and that of the university sector is very small. Roughly two-thirds of total R&D funds are spent by private enterprises. The shares of GRIs and universities are 15 and 10%, respectively. National/public universities rely mostly on government funds, while the private universities tend to be self-financing.

3. R&D *activities and international competitiveness*

R&D resources tend to be concentrated in a small number of industries, especially the ICT (*e.g.* communications equipment, semiconductors, computers, and electrical and electronic products) and automobile sectors, while the sources of international competitiveness are spread over a diverse set of industries, including some with low research intensities. Some R&D-intensive industries (*e.g.* precision instruments and pharmaceuticals) have been lagging in improving their international competitiveness, in contrast with some scale-intensive sectors (*e.g.* plastic and rubber, shipbuilding, and iron and steel) and labor-intensive sectors (*e.g.* textiles and leather).

This implies that Korean industry has not yet fully harnessed the potential of R&D to add greater value to their products. While Korea is a major exporter of high-technology products, the value content of its exports, including high-technology products, remains low. For instance, its up-market share in exports to EU-15 countries is below the OECD average, while its down-market share ranks fourth, after Turkey, the Czech Republic and Poland.[2]

C. Main features and weaknesses of the Korean innovation system

1. *Weakness of the basic knowledge effort*

In terms of basic research, Korea shows two contrasting features: it ranks in sixth place among the OECD countries in overall spending on basic research as a percentage of GDP; but it has one of the lowest shares of basic research in all R&D activities (Figure 5.1). The favorable position in terms of spending on basic research reflects the fact that Korean business enterprises perform more basic research as a share of GDP than do firms in any other OECD country (with the exception of Switzerland). However, the low overall basic research share indicates that public research sectors, including the GRIs and the universities, spend relatively less on basic research than in most other OECD countries.[3]

Korea's universities are characterized as being weak in research and are mainly oriented toward general education. Their share in total R&D expenditures is one of the lowest in the OECD area (Figure 5.2). To a large extent, this is due to high student-teacher ratios and heavy teaching loads, inadequate research infrastructure (*e.g.* a lack of experimental facilities), and a lack of qualified research manpower and support staff. However, the shortage of human resources is not sufficient to explain Korea's poor performance as measured in terms of scientific publications and patents. The ratio of R&D expenditures to researchers, a proxy for research infrastructure, is almost equivalent to that of most OECD countries. This points to the existence of institutional problems and a lack of incentives for

Figure 5.1. **Basic research as a percentage of GDP by sector of performance, 1997**

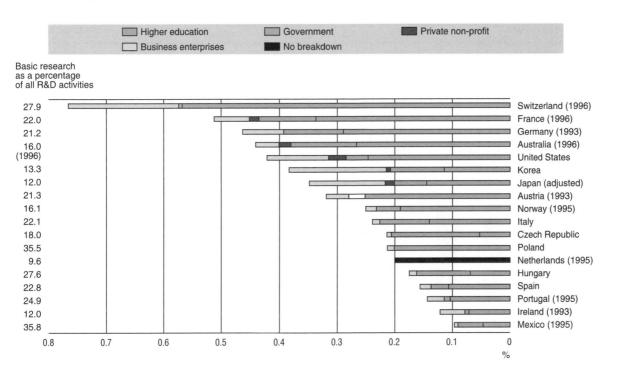

Source: OECD (1999b).

research in the higher education sector. Until very recently, research had not been a primary concern in most Korean universities and quality of research is not given much weight in recruitment and promotion.

In addition, Korean universities do not specialize in basic research. Their participation in wider R&D activities may suggest better relationships with other innovation actors. However, this creates a serious bias in the orientation of university research and weakens the domestic knowledge base. The problem is closely related to flows of R&D funds. Compared with the GRIs, public funding for university research in general is low, and there is a sharp contrast between funding for the national and public universities and funding for private universities. Public research funding for the latter stands at half the level of that for national and public universities. Consequently, universities tend to seek other sources of funding, notably from private enterprises, which may also serve to bias university research away from enhancing the knowledge base.

2. *Private sector R&D*

The limitations of internalization in large companies

Private enterprises have endeavored to upgrade their technological capabilities by establishing in-house research labs; they have rapidly increased their R&D expenditures since the early 1980s, with large companies, notably the *chaebol*, taking the lead. This process called for intensive assimilation and adaptation, and led to the successful development of production capacities for standardized products using mature technologies. Until recently, most R&D laboratories were organized and managed in the same way as other areas of business, *i.e.* in the framework of centralized and hierarchical structures that facilitate the mobilization of existing financial and human resources into new business lines while utiliz-

Figure 5.2. **Higher education expenditure on R&D per researcher**

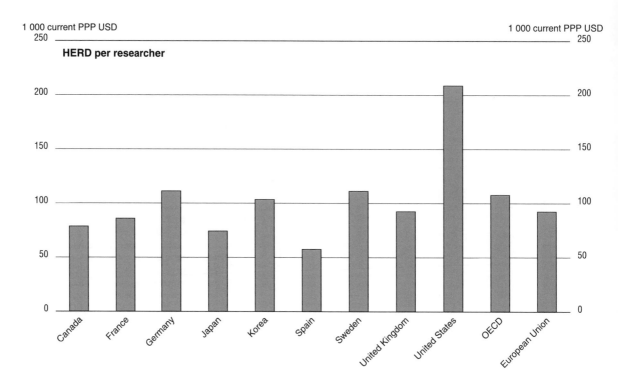

Source: Calculated from OECD (1999*b*).

ing the know-how and production inputs of existing ones. Combined with lower labor costs, these management strategies, including in the R&D area, contributed to the strength of the Korean conglomerates compared to their foreign competitors.

The weakness of current R&D system is that large research labs with centralized and hierarchical structures are inherently lacking in flexibility; in addition, the cost of dismantling existing organisational structures to meet the needs of the knowledge-based economy could prove prohibitive. A further limitation of excessive internalization is that it weakens the need for closer co-operation with other actors in the innovation system. A vicious circle of self-fulfilling internal ties is therefore blocking the further development of the KIS.

The structural imbalance between large companies and SMEs

Over the last decade, the share of the top-20 companies in total business enterprise R&D has increased steadily. In contrast, there has been no substantial increase in the R&D intensity of SMEs. This imbalance between the significant internal R&D effort by large companies and the lack of R&D spending by technologically weak SMEs, raises a serious problem for the KIS. As of 1995, only 0.7% of manufacturing enterprises with fewer than 100 employees performed R&D, compared to 19.1% of enterprises with between 100 and 299 employees and two-thirds of enterprises with more than 299 employees. In the KBE, industrial and production activities become technologically more demanding and complex, and the existence of supporting industries – where SMEs play a critical role – is a key factor in a company's international competitiveness as well as in the development of the producer-user interactions that are an essential source of innovation and technology diffusion.

3. Government policies and programs

The effectiveness of government R&D support policy

Korea has implemented a number of policy measures targeted at promoting technological innovation by private enterprises. The government's R&D support system includes: *i*) tax incentives; *ii*) financial incentives; *iii*) procurement; *iv*) technical information support; *v*) human resource support; *vi*) promotion of co-operative research; *vii*) technology support for SMEs; *viii*) support for the commercialization of new technologies; and *ix*) promotion of the establishment of research labs. Each of these categories contains detailed policy measures (OECD, 1996*a*).

Given the patchiness of the evidence, the effectiveness of these policy categories and measures is hard to assess objectively, although recent policy measures to increase government funding for basic R&D and enhance policy co-ordination by the NSTC are cited as good policy practices (OECD, 1999*c*). However, some important issues merit consideration. The Korea Industrial Technology Association's *Industrial Technology White Paper* 1998 points out that the absolute amount of government financial support is too low to have a substantial effect on innovation by private enterprises (KITA, 1998). Other reports express the need for increased government support.[4] The appropriate scale of government R&D support is an issue that deserves an in-depth and comprehensive study. Yet, prior to increasing government support, it is important to evaluate the effectiveness of current policy measures and thereby allocate resources more efficiently.[5] A recent STEPI report covering seven tax incentive measures and six other measures concludes that these measures are only marginally helpful in stimulating business enterprises' innovation activities. The report emphasizes the need for streamlining and restructuring support measures which are overly complex (Song and Shin, 1998).

Notwithstanding numerous government policy measures, the technological capabilities of Korean SMEs are still lagging. Among the OECD countries, Korea has the lowest share of SMEs engaged in R&D activities, despite the fact that the share of these enterprises in total government funding of business R&D is relatively high.[6] This raises questions concerning the appropriateness of policy measures to foster R&D-related innovations in Korean SMEs.

The STEPI report mentioned above indicates that many technology support measures, including the technology reserve fund system, are not particularly effective in increasing innovativeness among SMEs. In contrast, measures such as the government procurement system and the technology transfer system from public research institutes and universities, appear to be quite successful in enhancing new product development by SMEs. Nonetheless, proportionally greater government funding is allocated to relatively ineffective measures.

National R&D programs (NRDP) *and the role of the GRIs*[7]

Korea's national R&D programs (NRDPs) were launched by MOST in 1982 as the "Special R&D Program (SRDP)". The implementation of the NRDPs is closely related to the role of the GRIs.[8] The NRDPs provided a momentous opportunity for revitalizing GRI research, *i.e.* as a complement to research areas that would not be covered by the private sector. There have been substantial achievements.[9] However, although the Korean Government has tried to balance the NRDPs' mission- and diffusion-oriented efforts, NRDPs remain biased towards mission-oriented R&D. Strategic targeting is also popular in programs of other OECD countries and in the EU framework programs. The issue is how to use public resources more efficiently and how to build diffusion mechanisms into the programs. Korea has tried to use NRDPs to support diffusion programs, and has created separate diffusion programs, but so far this strategy has not proved very effective.

The rationale for the NRDPs and the *raison d'être* of the GRIs are to support and complement industry research by conducting upstream research that would not be performed by industry alone. However, most GRIs perform downstream research and experimental development, with several implications: *i*) a lack of long-term basic research, which weakens the basic science/knowledge base of the KIS; and *ii*) research tends to overlap with that of the private sector, leading to duplication and the crowding out of resources.

The weakness of the science/knowledge base of the KIS is clearly related to the general orientation of the NRDPs. Owing to the concentration of public R&D resources on GRIs, the universities play a minor role in the national innovation system. In 1998, 65% of NRDP resources went to the GRIs as main contractors, compared with only 9 and 2%, respectively, to the universities and industry. University researchers participate in most NRDPs but, generally, to a minor extent. Given that universities have 58% of PhDs in full-time equivalent terms (75% in absolute numbers), the contract pattern and budget allocation of the NRDPs are strongly biased against the universities.

The problem of overlap with the private sector is severe. A STEPI report, based on a 1995 survey, noted that more than half of the companies participating in the NRDPs would have undertaken the project even in the absence of government support.[10] Reasons for this include: *i*) the NRDPs are not sufficiently clearly planned, do not reflect industry needs and do not identify elements lacking in private research; and *ii*) the majority of the private partners of NRDPs are large companies, notably *chaebol*, which are sufficiently mature to implement the most commercially oriented NRDP research topics themselves. The government has recently planned a new R&D program, based on long-term technology forecasting and a survey of industry needs.

The NRDPs as a whole have been unsatisfactory in terms of utilizing national R&D results (see Box 5.1). There are several reasons for this, including the lack of policy co-ordination. Several measures, including the establishment of the National S&T Council (NSTC), aimed at harmonizing individual ministries' R&D programs, have been implemented to improve results. Other recent measures include the enactment of the Special Law for S&T Innovation and programs to improve technology transfer and the establishment of venture businesses to commercialize research results and set up technology business incubators in GRIs.

Several NRDPs have targeted SMEs, but with disappointing results. From 1983 to 1997, when MOST introduced the "New Technology Commercialization Program" for SMEs, the total amount spent for the

---

Box 5.1. The under-utilization of national R&D results – STEPI report

Based on surveys and interviews with 947 principal researchers who had participated in NRDPs in the past three years, including those of MOST and other ministries, the report analyses how effectively R&D results from NRDPs are utilized. First, diffusion from GRIs to private enterprises is moderately successful, whereas diffusion from universities is less satisfactory. Second, a number of bottlenecks impede diffusion: 42% of GRI and university respondents pointed to the lack of technological capability of private enterprises; in contrast, 40% of private enterprise respondents point out the lack of GRI and university technological know-how and knowledge for solving private firms' technical problems. Third, public R&D institutions, including universities, responded that private firms are not interested in technology diffusion from them; they also emphasized the lack of absorptive capacity in private firms. Private firms responded that public R&D institutions have not made sufficient efforts to increase public awareness of national R&D projects. Fourth, the best way to diffuse new technological knowledge generated by GRIs is to transfer technical personnel trained in GRIs to private firms. Due to their heavy teaching load, university professors are not actively involved in technology diffusion. GRI researchers have not efficiently extended their R&D results to commercialization. Moreover, no professional organisation exists to effectively link universities and GRIs to private firms. The report concludes that all these factors contribute to the under-utilization of national R&D results in Korea.

Source: Oh (1997).

"SME Support Program", one of ten NRDP categories, represented only 2.4% of the overall budget for NRDPs. In most other NRDPs, SME participation has been negligible.

4.   Weak global linkages

Knowledge and technology flow across borders through various channels. Korea's traditional ways of acquiring global knowledge and technology sources were through arm's-length licensing and "brain gain" – the return of Korean scientists and engineers from abroad. Although useful, these two methods have limitations. Licensed technologies are protected and tend to be mature, meaning that the potential for further innovation is relatively low. According to a report by the Korean Development Bank, most technologies licensed to Korea are in the mature stages of their life cycle (Korea Development Bank, 1991). Moreover in recent years, the number of Korean students studying abroad has decreased markedly,[11] suggesting that the benefits of the "brain gain" will decrease as time goes on.

Other means of knowledge and technology transfer have not been used extensively; for instance, until the financial crisis FDI played a very minor role and joint ventures were not popular. Consequently, Korea did not take advantage of knowledge and technology inflows from multinational activities.[12] In the areas of science and technology research, Korea has weak global linkages: compared to other OECD countries, the country ranks very low in cross-border ownership of inventions, cross-border co-authorship of scientific articles and co-invention of patents.

D.   The new policy agenda

1.   The need to adapt STI policy

It is clear from the above that Korean science, technology and innovation policy needs to be reoriented. A number of issues were identified in the 1995 OECD policy review (OECD, 1996a), and some policy measures have been taken.[13] However, the weaknesses of the innovation system have become more apparent in the wake of the recent financial crisis, and reforms can no longer be postponed.

Korea needs to reorient its R&D efforts, including the way in which they are managed and implemented by the public and private sectors. University research should be strengthened and the private sector, and particularly the chaebol, should be encouraged to work in partnership with other actors in the innovation process. The technological and organisational capabilities of SMEs need upgrading. The role of the GRIs needs to be redefined to place greater emphasis on basic research to meet future demand, in addition to performing research on technologies that serve the public interest such as environment, space and nuclear energy, and which do not receive enough attention by the private sector.

Box 5.2 presents a summary overview of the changes required of the different actors in the transition from the "catch-up" model to the "KBE" model. Considering Korea's current level of development, it can still benefit from many elements of the catch-up model, but it needs to actively begin the transition to the KBE model. This chapter provides a detailed overview of the actions that will be required of government.

The Korean Government has elaborated a "Long-term Vision for S&T Development towards the Year 2025" which sets new guidelines for its S&T strategy; based on the transition: i) from a government-led, development-centered, domestically networked R&D system to a private-led, distribution-centered, globally networked R&D system; ii) from a supply expansion policy to an efficient utilization policy; and iii) from a policy meeting short-term demands to a policy with longer-term perspectives that creates new markets (MOST, 2000). These broad guidelines provide a framework for the new policy approaches needed for the transition from the catch-up model to the KBE model.

The nature of the innovation process, including its linkages with more basic research activities, calls for continuous rethinking. Competitive markets stimulate innovation and allow firms and individuals to derive the benefits from knowledge accumulation. At the same time, firms are not simply production organisations but also learning organisations, whose efficiency depends on a set of country-specific institutional, infrastructural and cultural conditions. In addition to correcting market failures (provision

---

Box 5.2.   KIS for the transition to a knowledge-based economy

	Catch-up model	KBE model	Requirements
System as a whole	• Compartmentalized • Centralized	• New engine of growth • Integrated • Participatory • Decentralized	• Build up an indigenous knowledge base • Strengthen domestic and international linkages • Promote a regional innovation system
Government	• Developmental • Client-oriented • Mission-oriented • Sectoral promotion	• Catalytic • Collaborative • Diffusion-oriented • Cluster approach	• Define a new role for government • Enhance inter-ministerial co-ordination and partnership with industry • Build diffusion mechanism in government R&D programs • Improve framework conditions
University	• General education orientation • Minor role as knowledge producer	• Higher research orientation • Primary source of new knowledge in domestic context • Main producer of *both* graduates and knowledge	• Adapt the education system to the needs of the market • Strengthen research capability, especially basic research
GRI	• Targeted technology development • Whole spectrum of R&D	• Higher contribution to knowledge base • Clear division of labor with university and industry	• Redefine the role of the GRIs • Realign the GRIs
Industry	• Rapid market expansion • Volume/cost advantage • Industrial/technological widening • Hierarchical production system	• Secure market specialization • Higher value-addition and economic rents via innovation • Industrial/technological deepening • More horizontal relationships	• Define a new business strategy • Capitalize on R&D resources • Redefine the role of the *chaebol* • Strengthen the technological capabilities of SMEs • Strengthen linkages and clusters

of public goods, intellectual property rights, subsidization of R&D), governments have a responsibility to improve the institutional framework for knowledge interactions among firms and between market and non-market organisations. In particular, most OECD countries are introducing new policy and institutional measures aimed at stimulating the diffusion of public R&D results throughout the economy (the Danish and Dutch programs are good examples of such measures).

The private sector can improve its R&D efficiency by reorganizing its R&D activities. Many companies have begun to move in this direction, particularly since the financial crisis, by: downsizing R&D labs; adopting more horizontal structures to create networked R&D units within and outside the company; encouraging spin-offs; and introducing performance-based remuneration systems for R&D personnel, etc. A very positive recent phenomenon is that specialized R&D laboratories are emerging as spin-offs from firms that are downsizing and from universities.

The efforts of private enterprises to reorganize their R&D activities have been triggered by the changes in the business environment that have come about in the aftermath of the crisis, especially those affecting corporate governance structures. Macroeconomic and competition policies will also have significant effects on R&D activities (see Chapter 2). This is why the OECD calls for government initiatives to ensure "framework conditions" (incentive and institutional regimes) that are conducive to innovation (OECD, 1996*b*; 1998*b*). Such conditions include taxation and accounting practices that affect firms' strategies with respect to technological and organisational assets and human resources, as well as regulations and public procurement practices that affect forms of organisation, for example, for co-operation among firms.

A key feature of a knowledge-based economy is that agglomeration economies at the regional level, network externalities and dynamic economies of scale in clusters of technologically related activities are important sources of increasing returns to private and public investment in R&D. As a consequence, most OECD countries are shifting their STI policy away from conventional sectoral promotion toward a cluster approach. Finally, the experiences of the most successful OECD countries demonstrate that increasing the efficiency of STI policy requires improved policy co-ordination and evaluation mechanisms (see Chapter 7).

1. *Enhancing basic research capabilities*

The extent to which a country should devote and allocate resources to basic research is always difficult to estimate. There are trade-offs as well as complementarities between basic research and applied research. The challenge for Korea is to make strategic use of existing knowledge bases, both domestic and foreign. This will require a comprehensive review of the nation's basic and applied research capabilities in all scientific fields *vis-à-vis* international standards. The evaluation should be carried out by teams made up of eminent foreign and Korean experts in concerned disciplines, and best practices should be adopted: this is key for ensuring sound judgements on the strengths and weaknesses of the current research system and for defining an overall research strategy which encompasses both the universities and the public research institutes.[14]

A central issue in enhancing Korea's knowledge base is the upgrading of university research. Korean universities have grown massively in recent years, with the number of doctoral degrees awarded in science and engineering increasing more than eleven-fold, from 160 in 1980 to 1 920 in 1995. Yet, the quality and quantity of research is insufficient. The government should play a catalytic role in improving this situation by: *i*) increasing public support in order to encourage basic research; and *ii*) making better use of the "centers of excellence" framework under which universities compete and specialize on the basis of research quality (*Brain Korea 21* is a step in the right direction and should perhaps be expanded);[15] and *iii*) designing a system whereby the universities would supply qualified graduates to industry. Other policy recommendations include:

- *Increased funding for the research base.* In most OECD countries, science policy is evolving toward greater emphasis on the technological and industrial applicability of scientific research. Underlying this change is a strong domestic science base underpinned by the research capabilities of universities or national research centers. Korea needs to focus on upgrading its basic science and research base. Some relevant policy measures have been adopted; for instance, the MOST's ERC/SRC/RRC programs and the Ministry of Education's Academic Promotion Fund. However, further secure public funding for more basic and long-term research is recommended.

- *Promoting centers of excellence.* Based on transparent reviews (with the participation of foreign experts), a small number of universities should strongly focus on research (as is the case in Australia and Canada). To date, KAIST and POSTECH have achieved this goal. OECD (1996a) concluded that MOST's SRC/ERC programs were successful in promoting research in universities, mainly due to the following factors: funding was dependent on research excellence; regular evaluations were performed; public support was based on long-term research; and management autonomy was secured. These factors, together with a relatively clearer mechanism of competition, have provided the "framework" for research in universities. This framework should be applied more generally to the allocation of research funds to universities.

- *Provision of specialists.* The supply of qualified technical and scientific graduates to industry represents the most direct contribution of the universities to the innovation system. In most successful education systems in OECD countries, university research and training are fundamental to (regional) innovation clusters. A number of European countries have developed excellent technical school systems (*e.g.* Austria's *Fachhochschulen*) or university curricula adapted to respond to business needs (*e.g.* University Colleges in Ireland). Some Korean universities are moving in this direction but, as noted in Chapter 3, the government needs to further deregulate the university sector in such a way as to promote greater adjustment by the universities to changing market needs.

2. *Redefining the role of the GRIs and their funding*

The Law for Establishment, Administration and Promotion of Government-funded Research Institutes (hereafter, GRI Law) was enacted in January 1999. Based on this law, the GRIs depending on each ministry were transferred to five Research Councils under the responsibility of the Prime Minister's Office. Science and Technology research institutes were transferred to three S&T Research Councils: Fundamental, Industrial and Public. The GRI Law and the accompanying restructuring process aim to induce administrative innovation, effective human resource management and enhanced productivity in research projects, although it is as yet too early to assess the results of the measures. As part of this restructuring process, the Korean Government needs to address a number of fundamental issues:

- *Redirecting* GRI *research toward more basic and long-term research through secure public funding.* Rapid increases in university and private sector research capabilities have occurred over the years. In response, the government has introduced several policy measures aimed at renewing the GRIs, but the task is not yet complete. For effective restructuring, the research orientation of the GRIs needs to change. This implies ensuring secure government funding for more basic and long-term research by the GRIs. Based on the experiences of OECD countries with similar laboratories, the ratio of outside funding to secure funds ranges from 35-65% to 55-45%, depending on the nature of disciplines or technologies in which they operate.

- *Redesigning funding mechanisms.* In addition to increasing public funding for basic and long-term research carried out by the GRIs, the funding mechanism needs to be strengthened. The criteria for allocating public funding (including clear guidelines on the economic rationale for such funding) as well as the process (*e.g.* peer review of competing proposals) should be strengthened. The government has instituted the principle of contractual research, through which industry's participation and linkages will be increased.[16] In 1996, the Project Base System was introduced as a national R&D management system to enhance research performance and competitiveness. The policy, by which government funding is made conditional on matching funds from the business sector for all R&D projects in which it is involved, is also a step in the right direction and should be actively pursued.

- *Repositioning the GRIs vis-à-vis other actors in the innovation process.* The current role of the GRIs overlaps with that of other innovation actors in universities and the private sector. In particular, it appears that the government uses the GRIs as agencies to undertake ministries' R&D programs. This strategy has serious drawbacks. For instance, the GRIs do not educate R&D personnel, a key function of public research institutions in many European countries. In a longer-term perspective, the Korean Government should consider how best to integrate the research capability of the GRIs with the universities' mission of education. The establishment of joint research teams or laboratories – with similar rules relating to evaluation, remuneration and promotion for all researchers involved – could be an efficient way of achieving the necessary integration (see the French experience with the "associated CNRS/university laboratories") (OECD, 1999c).

- *Ensuring long-term consistent policy design.* The GRIs play a pivotal role in orchestrating nation-wide innovation activities. The government needs to establish a long-term plan for the GRIs within the KIS; this could involve creating a GRI Realignment Commission, composed equally of representatives from government, industry, university and GRIs. It would be desirable for such a Commission to have an advisory board composed of eminent foreign experts, including Korean expatriates. Critical institutional redesign factors would include greater autonomy, mission rationalization, institutionalization of co-ordinating mechanisms and job retraining and relocation programs for displaced employees.

3. Strengthening knowledge diffusion and system linkages

Technology policies in OECD countries are converging toward two overriding objectives: *i)* complementing market forces where this would yield the highest social return (rather than allocating public support according to predefined sectoral or political priorities); and *ii)* fostering the development of linkages among all actors in the innovation system and providing these actors with market-compatible incentives. In particular, policy measures are shifting away from conventional subsidization towards public-private partnerships in which co-operation at the design stage and risk/cost sharing at the implementation stage ensure that the economic value of R&D activities is improved. For "mission-oriented" programs to be effective, a systemic approach is needed that provides a framework for a more market-driven, bottom-up definition of objectives and more decentralized implementation.[17] Policy implications for Korea follow:

- *Greater emphasis on the diffusion of* NRDPs. As mentioned above, the R&D results of both mission- and diffusion-oriented NRDPs are poorly diffused and utilized. The government should, in consultation with all the parties concerned, promote the commercialization of NRDPs' R&D results through public-private partnerships. Korean policies tend to target direct support measures, with other measures attracting less attention. In contrast, many OECD countries direct their support to other areas, such as promoting networking and clustering (*e.g.* the Dutch Clustering policies and the German BioRegio Initiative), expanding SME access to the public research system (*e.g.* the United States' Small Business Technology Transfer Program), etc. (OECD, 1999*d*).

- *Bridging institutions.* In addition to placing greater emphasis on the diffusion of NRDPs, it is very important to institutionalize the diffusion mechanisms. This can be done in two ways: *i)* provide incentives for existing research institutions to induce them to increase their diffusion efforts; and *ii)* establish intermediary bridging institutions. Korea is already moving in the first direction, but the second also deserves consideration. For instance, despite its many achievements, Taeduck Science Town has been often criticized for its weak industrial linkages. Government can either induce the GRIs in the town to increase their efforts or establish a specialized institution with that specific objective. A good example is Chinese Taipei's Hsinchu Science-based Industrial Park.

- *Intermediary institutions in the private sector.* Intermediary institutions can serve to strengthen the linkages between public and higher-education research and industry. Korea lacks the university-industry interface units and technology centers which are popular in other OECD countries. The high concentration of private R&D in the *chaebol* has hindered the development of the independent service firms which play a significant role in the diffusion of knowledge in advanced countries and allow more efficient outsourcing of technological, management and organisational services. The specialized research labs which have been spun off from large firms in the wake of the financial crisis offer a welcome opportunity to fill some of the missing links in the KIS. Government should encourage their growth, for instance by allowing them access to the public research system and results.

- *Strengthening global linkages.* Korea's scientific and technological activities need to be better integrated into global networks. As of 1998, 19 Korean journals featured among the 5 600 journals covered by the Science Citation Index (SCI). The Korean Government should seek to create a more internationalized science community, for instance by encouraging the enrolment of capable foreign scholars in Korean universities and public research institutes through grants and fellowships. Government and public research institutes could establish an Advisory Board composed of eminent foreign scholars and Korean expatriates. The restructuring of the *chaebol* and the surge of foreign investment provide an opportunity to diversify the sources of knowledge acquisition. Government should seek to link recent inflows of foreign investment to domestic R&D activities. The integration of Korea into international innovation networks will require initiatives in other areas, such as international technology co-operation, including international business alliances, scholarships, etc.

4. Increasing the mobility of human resources in science and technology

Technical progress and the move toward the knowledge-based economy are increasing the demand for skilled labor and spurring an upgrading of skills. This raises several important policy challenges for governments, one of which is increasing human mobility. The movement of science and technology personnel between sectors, large and small firms, and across national borders is an important conduit for technology transfer. The Korean Government should strive to implement policy initiatives aimed at increasing human mobility in the KIS. These should focus on:

- *Removing (regulatory) obstacles that impair the mobility of researchers between the public and private sectors.* At a general level, the barriers and obstacles to labor market mobility for S&T personnel relate to regulations on employment (hiring and firing), pension rules and wage bargaining arrangements. Regulations such as employment protection legislation can act as a barrier to flexibility and mobility. In the public sector, researchers may not be willing to abandon permanent employment for employment in industry, even at a higher wage. Other factors that hinder mobility include age limits on junior faculty or research posts. In all these areas, government needs to make further efforts.

- *Adopting more open recruitment policies in high-level government positions with responsibilities in policy making and implementation.* The current official recruitment system creates an invisible wall between government and business. Government should encourage much wider participation of business in NRDPs, not only as research partners but also as key players in program design and evaluation.

5. Stimulating innovation in SMEs

In line with streamlining its R&D support system, the Korean Government should make greater efforts to identify the most pressing needs of SMEs. First, awareness of the importance of innovation is still low among SMEs, particularly among the smaller ones; government should increase information provision services for SMEs. Second, SME policies should aim at enhancing the absorptive capacity of SMEs through the provision of qualified R&D personnel and at promoting technology transfer from public research institutions. Third, government should comprehensively review support measures and reallocate public funds in consequence. In accordance with this comprehensive review and reallocation, implementation procedures need to be re-examined in light of international best practices (*e.g.* the US Small Business Innovation Research Program and Small Business Technology Transfer Program).

The system for supporting innovation in SMEs needs: *i)* increased resources (both financial and human); and *ii)* a regional/local base. In this connection, the experiences of the European countries may provide useful insights. For example, the French Innovation Agency (ANVAR) provides support to R&D and innovation in SMEs, with a total yearly budget of USD 200 million, and with the assistance of 24 regional offices which enjoy full decision-making power of up to USD 20 000 per project. Assistance measures include direct support to innovation – reimbursable in case of success, subsidies to employment of scientists by SMEs, support for collaborative projects with a university or a public laboratory, seed money for enterprise creation, etc.

A network of local service points providing technical information and assistance to SMEs should be set up. From this viewpoint, the Korean system, based on the "Technical Academy", is rather undeveloped, with very limited resources and practically no local "antennas". Korea does not have the capacity to build up a large infrastructure, such as the Japanese prefectural laboratories network, but should aim for lighter yet efficient structures, similar to the Danish system (OECD, 1995).

The Korean Government should strive to expand SME access to the public research system. To date, large enterprises have been the main partners of the NRDPs and SME participation should be encouraged. The government should also earmark a set percentage (up to 10%) of the R&D procurement of each ministry to SMEs, in line with the successful US experience. Further, knowledge diffusion should be facilitated through networks, which can be both horizontal (*e.g.* with SMEs setting up a research consortium with public research institutes, including universities) and vertical (*e.g.* with the *chaebol* acting as the main organizers through their SME subcontractors).

In recognition of the importance of venture capital in creating economic and employment benefits and supporting the start-up of technology-based SMEs, the Korean Government has recently initiated several policy measures. In 1997, the Venture Business Promotion Law[18] was enacted and financial support for venture business has been substantially increased over the last two years. These policy initiatives appear to be contributing significantly to the creation of an entrepreneurial environment that favors risk-taking. The government now needs to focus more on the overall regulatory and supervisory environment for venture capital to ensure that there is proper discipline and transparency in this booming market.[19]

6. *Shifting from sectoral support to cluster promotion*

Cluster-based innovation and technology policy offers many advantages and the cluster approach has become a key policy tool in a number of OECD countries. Clusters can systematically enhance linkages among innovation activities, thereby maximizing the value added of production activities. Private enterprises – and in particular SMEs – can efficiently draw upon the knowledge-creating institutions such as universities and research labs which form the hub of the cluster. Governments can promote clusters in several direct and indirect ways; their cluster policies are closely related to regional development policies and can help to reduce regional imbalances (see Box 5.3). The cluster approach has several implications for Korea.

Traditionally, the *chaebol* have played the role of clustering institutions through subsidiary and subcontracting companies, while government has pursued a sectoral support policy. This division of labor is becoming less and less viable; it has led to an over-expansion of the *chaebol* system while impeding the exploitation of intersectoral synergies in the innovation process. The Korean Government should facilitate networking and clustering, in particular by: *i)* establishing a more competitive environment through competition policy; *ii)* providing appropriate infrastructures and incentives, in particular for SMEs; and *iii)* improving co-ordination of regional and national policies.

Box 5.3. **From a sectoral to a cluster approach: innovation and technology policy in the knowledge-based economy**

Clusters are networks of interdependent firms, knowledge-producing institutions (universities, research institutes, technology-providing firms), bridging institutions (*e.g.* providers of technical or consultancy services) and customers, linked in a value-added-creating production chain. The concept of clusters goes beyond that of firm networks and captures all forms of knowledge sharing and exchange. The analysis of clusters also goes beyond traditional sectoral analysis, as it accounts for the interconnection of firms outside their sectoral boundaries. In many OECD countries, clusters are seen as the drivers of growth and employment. Governments can nurture the development of innovative clusters primarily through regional and local policies and development programs, and by providing the appropriate policy frameworks in areas such as education, finance, competition and regulation. Some best practices include:

- Creating a platform for government's dialogue with the business sector (the Netherlands).
- Focusing R&D schemes, innovative public procurement, investment incentives and the creation of "centers of excellence" (Sweden).
- Competing for government funding to provide incentives for firm networks to organize themselves on a regional basis (Germany).

Source: OECD (1999*c*).

7. Stimulating regional and local innovation initiatives

As clearly shown by the above considerations and abundantly illustrated by the experiences of the OECD countries, the regional and local dimension is key to the development of efficient innovation policies. Korea suffers from an administrative framework that does not facilitate spontaneous initiatives by regional and municipal authorities. The lack of financial autonomy of these authorities, their budget being fed by limited local taxes and funds provided by the central ministries, serves to exacerbate this inflexibility. However, some opportunities exist that could usefully be exploited by some of the entities – the six metropolises (including Seoul, Pusan, Kwangju) and the nine provincial governments – which do enjoy a significant budget. Central government can stimulate this move by providing matching funds for the establishment and overheads of the infrastructures required for supporting innovation in SMEs and constituting regional clusters. The experiences of some Nordic countries (*e.g.* Denmark and Finland) could be useful in this respect, as well as that of France (although the regional innovation initiatives benefited from an overall decentralization policy since the early 1980s which significantly increased the financial autonomy of the regions).

8. Improving co-ordination and evaluation at the central level

To be successful, innovation policy requires strong co-ordination between the various ministries and agencies involved in the innovation process. There is a real need for co-ordination between education policy (MOE), S&T policy (MOST) and industrial policy (MOCIE). The recent creation of the National S&T Council (NSTC) will allow for a more co-operative approach, and, if appropriately managed, could become an efficient instrument of change, facilitating the implementation of the necessary reforms. A good example in this area is the Finnish S&T Policy Council, which brings together government bodies and high-level industry representatives under the chairmanship of the Prime Minister, and has been instrumental in stimulating change in the Finnish innovation system.

In order to conduct their task efficiently, the central authorities, together with the above-mentioned NSTC, should commission and make use of appropriate evaluations. Some organisations, such as the Science and Technology Policy Institute (STEPI) and the Korea Institute for S&T Evaluation and Planning (KISTEP), have conducted rigorous and critical evaluations of government programs (see Section C3), and good use should be made of such monitoring capabilities. An international review of Korea's basic research competencies could usefully be performed and, as a complement to such a review, a technology foresight study would be instrumental in helping to place Korea's efforts in the broader international context. To date, Korea has not made much use of such benchmarking techniques, although they provide a useful tool for assessing global trends and highlighting areas where Korea should concentrate its efforts. The exercise should draw on the experiences of the scientific and business communities.[20]

E. Conclusions

The challenges facing the KIS call for important changes. As a number of these changes concern, directly or indirectly, the wider economic context, the adaptation of the KIS requires obtaining the commitment of the various actors in the innovation system. This is particularly so as, since the financial crisis, the government, industry and the research community have been engaged in a painful program of reforms. Some are positive, others need to be more carefully designed. However, the government has to redefine more carefully the mechanisms and means through which it can most effectively support the role of the different actors in the innovation system and the rationale for its subsidy and grant programs. This includes budgetary measures (increased funding, reallocation of resources, changes in funding methods, etc.), as well as new forms of incentives (particularly to facilitate linkages). It will also require institutional initiatives, such as the removal of regulations which hinder innovation (*e.g.* in universities), the creation of appropriate commissions (*e.g.* to "realign" the GRIs), or the introduction of targeted reviews and foresights. The implementation and co-ordination of the necessary measures relies mainly on the political will for reform, and is not a matter of lack of understanding, resources or administrative tools, as will be elaborated in the last chapter.

Notes

1. See Box 1.2 in Chapter 1 for a definition of a national innovation system.

2. Up-markets are where the unit values of exports are 15% above average, whereas down-markets are those where unit values of exports are 15% below average. For more information, see OECD (1999a).

3. For a comparison with other OECD countries, see OECD (1999a), Section 3.3 and appended tables.

4. For instance, according to Lee (1998), Korea uses only one-tenth of the possible public support measures allowed under the WTO subsidy rule.

5. KITA (1998) summarises the problems associated with government support: i) mismatches between the objectives of government support measures and industrial needs; ii) lack of complementarity between financial measures and tax incentives; iii) difficulty for SMEs in obtaining loans; iv) limitations on the mobilisation of funds through capital markets; v) high interest rates on bank loans; vi) an underdeveloped venture capital system; vii) inadequate banking regulations and practices. In particular, the *White Paper* points to the problem of overlapping policy measures of various ministries.

6. For a comparison with other OECD countries, see OECD (1999a), Section 5.4.

7. For a general explanation of national R&D programs, see OECD (1996a), Part I, Chapter VI.

8. Until 1982, a substantial part of the GRI budget relied on contract research from industry. However, as industries began to establish in-house research laboratories, industry's needs decreased. In response to this and other changes, the government restructured the GRIs.

9. For instance, the successful development of successive generations of DRAM was made possible by NRDPs co-ordinated by the ETRI GRI, with active participation by private enterprises such as Samsung, Hyundai and Gold Star. In terms of "output", it is reported that, by 1997, 687 items had been successfully commercialised; 482 patents granted in foreign countries; and 4 126 scientific papers published in international journals.

10. Hwang *et al.* (1997). Based on the survey of researchers who participated in NRDPs, the report highlights a number of other problems: a lack of differenciation among the programs; a significant bias towards commercial technology; a short-sighted time horizon for R&D; a discrepancy between R&D objectives and R&D performance/impacts; and inconsistencies in R&D policies and ineffective program management.

11. According to National Science Foundation (1999), the number of Korean science and engineering PhDs awarded in US universities peaked at 1 143 in 1994, and has since rapidly decreased.

12. Some OECD countries show positive evidence on the role of FDI in upgrading indigenous technological capability (Meyler, 1998). Djankov and Hoekman (1999) show that FDI in the Czech Republic had a great positive impact on total factor productivity in Czech enterprises. The same can be said of the experience of Singapore and Ireland.

13. In collaboration with the Korean Government, the OECD published the *Review of National Science and Technology Policy: Republic of Korea* in 1996 (OECD, 1996a). Its main (policy) recommendations have been widely circulated and referred to in Korea. These recommendations remain largely valid, and the Korean Government has recently initiated several policy changes based upon them. This section tries to der ive a more specific policy agenda in the context of a NIS-oriented, KBE perspective.

14. Sweden is very instructive in this respect, especially in terms of the experience of the Natural Sciences Research Council which conducts periodic international reviews of the research activities it supports in all fields (physics, chemistry, earth sciences, etc.). The system is remarkably expedient, efficient in terms of cost/quality ratio and also transparent. For further details, see OECD (1997).

15. *Brain Korea* 21, introduced in June 1999, aims to: create world-class graduate schools and leading regional universities; improve graduate school infrastructure; enhance research of graduate schools, balance the focus of academic studies; and enhance international co-operation by offering short- and long-term training for students and extending invitations to foreign professors.

16. European countries are extensively turning to contractual research to enhance the public research system. For instance, in the case of TNO in the Netherlands, the share of government funding in total turnover in 1997 was about 43%, with the rest coming mainly from contractual research.

17. For more detailed explanations, see OECD (1998*b*).

18. This law has been revised several times since its implementation, the latest change being made in January 2000. Some of the new provisions include: support for financing, location, technology development, human resource development, collaboration with research centers and access to the technology guarantee fund.

19. As noted by the European Venture Capital Association (1998, p. 11): "The most desirable government programs are those that strengthen the private venture capital sector and then, as private markets mature, are phased out. The economic and social benefits of such programs continue long after the government's direct role has ended."

20. Some OECD regions, for instance Ireland and Scotland, are more frequently including foreign companies in their industry-academia-government forums.

References

Djankov, Simeon, and Bernard Hoekman (1997),
"Foreign Investment and Productivity Growth in Czech Enterprises", *Policy Research Working Paper*, No. 2115, Washington, DC: World Bank.

European Venture Capital Association (1998),
White Paper: Priorities for Private Equity – Realising Europe's Entrepreneurial Potential, Brussels: EVCA.

Hwang, Yongsoo, *et al.* (1997),
"An Assessment of Government R&D Programs", Seoul: Science and Technology Policy Institute (in Korean).

Korea Development Bank (1991),
"Analysis of the Effects of Technology Imports", Seoul: KDB (in Korean).

Korea Industrial Technology Association (1998),
Industrial Technology White Paper 1998, Seoul: KITA (in Korean).

Lee, Won-Young (1998),
"Proposal for Improving Tax and Financial Incentive Systems", Seoul: Science and Technology Policy Institute (in Korean).

Meyler, Aidan (1998),
"Technology and Foreign Direct Investment in Ireland", Technical Paper No. 98/10, Economics Department, Trinity College, Dublin, Ireland.

Ministry of Science and Technology (Korea) (1997),
"Thirty Year History of Science and Technology", Seoul: MOST (in Korean).

Ministry of Science and Technology (Korea) (1998),
"Report on the Survey of R&D in Science and Technology", Seoul: MOST.

Ministry of Science and Technology (Korea) (1999),
"R&D Budget Statistics", Seoul: MOST.

Ministry of Science and Technology (Korea) (2000),
"Dream, Opportunity and Challenge of S&T Toward the Year 2025", Gwachon: MOST (in Korean).

National Science Foundation (1999),
"Science and Engineering Indicators", Arlington, Virginia: National Science Foundation.

OECD (1995),
Science, Technology and Innovation Policies: Denmark, Paris: OECD.

OECD (1996a),
Reviews of National Science and Technology Policy: Republic of Korea, Paris: OECD.

OECD (1996b),
Technology, Productivity and Job Creation, Vol. 2. Analytical Report, Paris: OECD.

OECD (1997),
"National Innovation Systems", free brochure, Paris: OECD.

OECD (1998a),
Science, Technology and Industry Outlook 1998, Paris: OECD.

OECD (1998b),
Technology, Productivity and Job Creation: Best Policy Practices, Paris: OECD.

OECD (1999a),
OECD *Science, Technology and Industry Scoreboard* 1999: *Benchmarking Knowledge-based Economies*, Paris: OECD.

OECD (1999b),
Main Science and Technology Indicators, Paris: OECD.

OECD (1999c),
"The Management of Science Systems", free brochure, Paris: OECD.

OECD (1999*d*),
   Managing National Innovation Systems, Paris: OECD.

OECD (2000),
   "Mobilising Human Resources for Innovation", free brochure, Paris: OECD.

Oh, Chai Kon (1997),
   "A Study on the Promotion of the Effective Diffusion of National R&D Results", Seoul: Science and Technology Policy Institute (in Korean).

Song, Wi-Chin, and Taeyoung Shin (1998),
   "Determinants of Success of New Technology Based Firms and Innovation Policy", Seoul: Science and Technology Policy Institute (in Korean).

US Patent and Trademark Office (1998),
   http://www.uspto.gov/web/offices/ac/ido/oeip/taf/reports.htm#CSTU.

Chapter 6

Promoting Knowledge-based Activities

A. Introduction

The level of development now attained by Korea, coupled with rising competition in high-volume, standard products and international technological and economic developments, make it essential for Korea to exploit knowledge effectively in economic activity. Part of the challenge is to raise its competitiveness in knowledge-based activities and industries – areas where a genuine capacity to innovate is vital. Over the past decades, Korea's industrial profile has moved toward what can be viewed as typical for an OECD country. Currently, knowledge-based industries, as defined by the OECD, account for 50.4% of value added in the business sector,[1] ranging from a low of 31.4% in Iceland to a high of 58.6% in Germany (former West Germany) (Figure 6.1).[2] These industries have in fact grown more rapidly than the average for the total business sector in practically all OECD countries (Figure 6.2). In Korea, knowledge-based industries account for 40.3% of business sector value added. Whereas knowledge-based manufacturing accounts for a relatively large share of GDP, the share of knowledge-based services is unusually small. On average, services now account for over 70% of economic activity in the OECD economies, compared to less than 20% for manufacturing (OECD, 1999a).

Furthermore, based on productivity and trade statistics, many knowledge-based industries seem to exhibit a quality gap relative to other industrialized countries (Woo, 1999). Table 6.1 summarizes Korea's position in a few knowledge-based industries, and indicates some key weaknesses and challenges.[3] As can be seen, Korea is well placed internationally in several of these industries, despite problems with respect to innovation capability, access to skilled personnel and value-added content. This situation reflects the Korean emphasis on mass production and price competition as means to expand market share, rather than on innovation and product differentiation to increase profitability through enhanced productivity and value added.

The government seeks to champion a triangular relationship in the Korean economy – one in which: i) the reformed *chaebol* would focus on their core competencies, becoming world-class, competitive conglomerates; and ii) knowledge-based high-technology SMEs and suppliers of parts would form a supporting pillar; while iii) foreign companies would invest in Korea, contributing with advanced manufacturing and management skills. These three functions are likely to be important for maintaining or developing competitiveness in the knowledge-based economy. More fundamental, however, is the task of strengthening the mechanisms for spontaneous development and use of knowledge in the economy. Throughout the OECD area, there is a similar ongoing shift in industrial policies away from defensive support of specific firms and industries toward building institutional conditions and policies which can strengthen the capacity of industry to respond to changing consumer demands and market opportunities.

It should be stressed that there is no static or absolute definition of knowledge-based industries or activities. Some tend to be more intensive than others in their use of knowledge, but the development and use of knowledge can be decisive for competitiveness in any economic activity. At the same time, historical circumstances, along with social and economic forces, shape industrial activities which, in turn, influence risks and opportunities for the future. Along with the policy areas presented in the previous chapters, at least five key sets of industry-related issues require special government action in Korea as the country seeks to undertake the transition to a knowledge-based economy: i) the dominating position and limited responsiveness of the *chaebol*; ii) the untapped potential of SMEs; iii) remaining barriers to con-

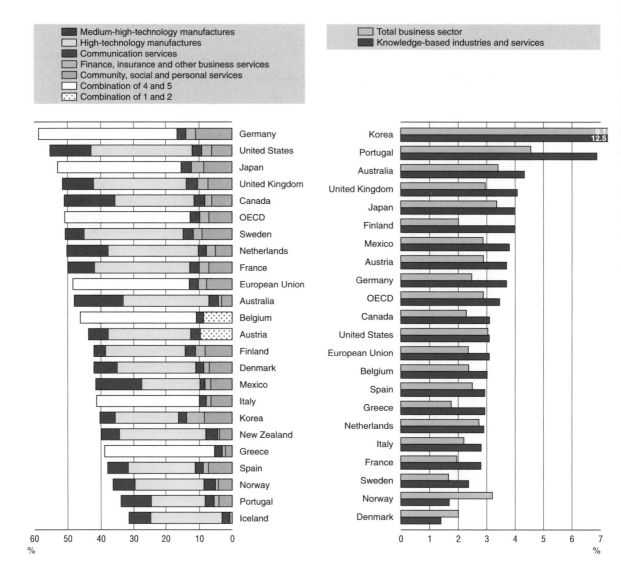

Figure 6.1. **Share of total business-sector value added by knowledge-based industries**

1996 or latest available year

Medium-high-technology manufactures
High-technology manufactures
Communication services
Finance, insurance and other business services
Community, social and personal services
Combination of 4 and 5
Combination of 1 and 2

Figure 6.2. **Real value-added growth by the total business sector and by knowledge-based industries**

compound average 1985-96 or closest years

Total business sector
Knowledge-based industries and services

Source: OECD (1999*a*).

tributions by foreign firms, including inward FDI; *iv*) impediments to high-value-added services; and *v*) inadequate incentives for firms to invest in intangible assets. Unless they are properly addressed, the presence of these issues will limit the pressures for change and the responsiveness of the private sector to new opportunities, and will thus serve to preserve outdated industrial structures at the expense of knowledge-based activities. The following sections discuss the policy issues that characterize these areas.

B. Reforming the *chaebol*

The prevailing industrial structure in Korea has been much influenced by the family-controlled conglomerates – the *chaebol*. Their investment patterns and their exploitation of scale economies have

Table 6.1. **Main characteristics of selected knowledge-intensive industries in Korea**

Industry	Strengths and position as of today	Weaknesses and challenges for further development
Semiconductor industry (SI), mainly semiconductor chips for a wide array of electronic devices	• Korean industry accounts for one-third of world production • The industry accounts for 10% of Korea's total exports (1997)	• Heavy dependence on a single product, *i.e.* memory chips, for nearly 90% of SI revenue • Heavy reliance on export markets for 90% of total production (1997) • Risk of losing competitiveness due to reduced investment in new product development and production capacity for future generations of products
Telecommunications equipment industry (TEI), including wire and radio communication equipment	• Production value USD 6 484 million • 58% of production sold in Korea, accounting for 63% of domestic market share (1997) • TEI production is anticipated to surge from 2000	• Increased competition in Korean market following Telecom and Information Technology Agreement (1997) • Competitiveness lags that of advanced countries • Insufficient innovation capability • Problems as regards core technology of TEI industry • Insufficient supply of high-quality human resources
Computer software and services (CSS), mainly packaged software industry and computer service industry	• CSS business turnover of USD 5 904 million (1997) • High (94%) market share in domestic market (1997)	• Reliance on imports for system software • Low export capability of the CSS industry, accounting for 1.6% of turnover (1997) • Lags the United States by an estimated 10-20 years in technological development • Insufficient supply of highly qualified software engineers • Difficulty for software industries to raise capital • Development of CSS hampered by weak protection of intellectual property rights
Fine chemicals industry (FCI), including pharmaceuticals, pesticides, cosmetics, dyes and organic pigments, paints and printing inks, and other high-value-added chemicals	• Production value of USD 17.4 billion • High (85%) market share in domestic market (1996) • Pharmaceuticals and cosmetics are twin engines of growth	• A high degree of reliance on imports for high-technology and high-value-added core inputs, *e.g.* 60-70% of intermediate input from abroad • Industry-related trade deficit of USD 2.9 billion in 1997 • Final products are mostly low-value-added common products • Lack of own technology, 56% of firms rely on technology imports (1997) • Low innovation capacity, *e.g.* ability to invent new drugs lags 50% below that of US firms
Biotechnology industry (BI), including the production of engineered biological products and all supporting businesses	• Production value of USD 892 million (1997) • 150 companies, of which two-thirds are side-line subsidiaries to large firms • National biotechnology development program (Biotech 2000) led by the government since 1993 • Large government budget awarded for R&D in biotechnology for USD 170 billion (1997)	• Korean firms possess some basic technology of BI thanks to R&D investment • Production technology mainly at the stage of imitation of advanced countries • Considerable gaps remain between scientific research, application research and technological commercialization • Anticipated difficulty in imitating advanced technology due to improved protection of intellectual property rights in the future
Environment industry (EI), including many diverse sectors, with environmental protection equipment production being the main component of the industry in Korea	• 11 700 firms, with sales revenue of USD 8.9 billion (1996) • Fast growth: 15% a year in 1990-97 • Large government environmental spending, at 1% of GDP during the 1990s • Policy environment conducive to EI, including the Environmental Vision 21 Policy, and Green Country Construction Plan • Government R&D support for the EI	• Insufficient technology capacity compared with firms from advanced countries, especially in the medium- and high-end product segments • The opening up of the Korean environmental protection market for equipment and services to foreign countries will be a major challenge to the Korean EI

Source: OECD (2000*a*).

been critical in establishing internationally competitive production in a number of capital-intensive industries:

- The conglomerates have effectively exploited economies to scale and devoted considerable resources to research.
- They have gained considerable experience in international markets and have established a strong competitive position in many product areas.

The rise of the *chaebol* during the 1960s and 1970s was supported by the Korean Government through preferential access to credit and protection from foreign competition. In a situation of weak financial markets and dependence on foreign technologies, the *chaebol* assumed the role of financial intermediaries. The diversity and size of their portfolios permitted them to spread and underwrite risk for investment in new or expanding areas. Their room for discretion provided them with the ability to act quickly – a factor that proved critical in a number of instances. Samsung, for example, became the world market leader in metal oxide semiconductors and the DRAM segment by the early 1990s, climbing from a market share of virtually zero in 1984. Hyundai and LG Semicon likewise emerged as the fourth and fifth largest producers of DRAMs in 1996. In shipbuilding, following an impressive campaign to increase capacity and sales, Korea now ranks number one, a position previously held by Japan.

The extensive diversification of the *chaebol* involved intense competition among themselves in numerous markets. In 1990, for example, Korea had 21 semiconductor producers, four shipbuilders and 14 television makers. Samsung chose to enter the automobile industry despite the fact that four other producers (Hyundai, Daewoo, Kia and Sangyong) were already supplying the relatively small domestic market. At the same time, the *chaebol* accumulated power and established modes of governance based on tight control by few actors, limited transparency, weak protection of minority shareholders and heavy dependence on bank lending rather than on equity. Despite reduced government support from the 1980s, the conglomerates have continued to account for a large proportion of economic activity.

The position, brand names, reputation and resources of the *chaebol* represent assets that can potentially be deployed to increase the production of knowledge-intensive goods and services. For instance, their strong competitive positions can be exploited to enable them to forge strategic alliances with foreign partners. On the other hand, the *chaebol* have generally lacked effective mechanisms for identifying – and disclosing – when operations go astray. There are "lock-in" effects along the trajectories of existing markets, products and technologies, and reduced incentives for innovation (see Chapter 5). This has contributed to excess capacity, extensive misallocation of resources and low profitability. Meanwhile, the reliance of the *chaebol* on imported technology and inputs to support expansion has often been paralleled by weak linkages with other parts of the Korean economy, including SMEs (OECD, 1999*b*; OECD, 2000*b*).

In the wake of the crisis, the government introduced measures to improve the global competitiveness of the *chaebol*. This has been done through: *i*) enhancement of market transparency; *ii*) elimination of cross debt guarantees; *iii*) capital structure improvement; *iv*) business consolidation into core competence areas; and *v*) improved management accountability. Additional supportive measures included: *i*) changes in the corporate governance of non-bank finance companies, many of which were controlled by *chaebol* and provided excessive funding to their affiliates; *ii*) curbing circulating investment and unfair intra-group transactions; and *iii*) blocking irregular means of inheritance. Some of the reforms are highlighted in Table 6.2.

The government's efforts toward restructuring have so far consisted in a mixture of market-oriented measures designed to liberalize and stimulate competition, and interventionist measures designed to force the *chaebol* to restructure, contract in areas of overcapacity and focus on their core competencies. Nevertheless, and despite increasing foreign competition, a web of vertically integrated structures and entrenched relationships continues to limit market contestability even in the face of extensive regulatory reform and apparent market openness. This raises a demand for continued action to enforce restructuring. At the same time, it must be clearly understood that the *chaebol* are strongly heterogeneous. In practice, these company groups display a variety of strengths and weaknesses, and there is no panacea for improved performance which can be universally applied by any public or private actor. The most effective instrument for the government to push for restructuring probably emanates from its role

Table 6.2.   **Measures to improve the general policy environment for business in Korea**

Area	Actions taken
Transparency	The *chaebol* are required to prepare consolidated financial statements, beginning in 1999, and quarterly reporting in 2000. These must include line-of-business reports as well as disclosure of intra-group transactions. Measures to increase the role and independence of external auditors have been introduced. The amendment of financial accounting standards in December 1998 has brought Korean practice into line with international standards.
Shareholder rights	The reform of the Commercial Code in December 1998 has enhanced the rights of minority shareholders by lowering ownership thresholds for initiating various types of adverse actions. The removal of "shadow-voting" in September 1998 permits investors to vote freely on such matters as mergers, business transfers and the election of directors and auditors.
Company directors	Listed companies are required to fill one-quarter of their boards of directors with outsiders, beginning in 1999. The responsibilities and obligations of directors have been clarified, while the definition of directors has been expanded to include any person who exercises a direct decision-making function in a corporation. Cumulative voting (which enhances minority shareholders' rights) for directors is now possible (but not required).
Intra-group relations	New debt guarantees between *chaebol* subsidiaries were prohibited in 1998; existing guarantees are to be eliminated by March 2000. The 25% ceiling on equity investments by *chaebol* subsidiaries in third companies (which had been intended to limit *chaebol* expansion) was lifted, with a view to facilitating restructuring. Holding companies are allowed, subject to restrictive conditions on debt and the structure of such companies.
Insolvency procedures	Re-organisation procedures (applicable to large enterprises that are deemed viable) have been strengthened through the introduction of economic criteria in evaluating applications, and through the establishment of time limits.
Investment	The ceiling on foreign shareholding in individual companies was abolished in 1998. All forms of mergers and acquisitions, including hostile take-overs by foreigners, are permitted. The number of business lines where foreign direct investment is restricted was reduced from 53 to 24.
Other	The ceiling on banks' equity investments in individual companies was temporarily boosted from 10 to 15%. As from 2000, the corporate tax system will not allow deductions for interest payments on any debt exceeding five times equity capital. A Corporate Restructuring Fund of KRW 16 trillion was created in October 1998 to assist SMEs.

Source: OECD (1999c).

as owner of a number of the major commercial banks: the state can – and should – ensure that they do not lend money unwisely. In addition, the Financial Supervisory Commission (FSC), as regulator of the banks, can take more affirmative action to ensure the health of financial institutions.

Beyond this, competition policy has an important role to play in ensuring effective competition. Indeed, as discussed in Chapter 2, the government has recently strengthened the position of the Korean Fair Trade Commission (KFTC). At the same time, with the amendment of the Fair Trade Act of 1986, the KFTC has been charged with regulating the activities of the *chaebol*. Restructuring the *chaebol* does not equate with antitrust. Monitoring the activities of the *chaebol* – 700 companies in all – is clearly a huge task. Since 1986, the main *chaebol* have grown, and experience shows that those that, at least until recently, have attracted the very highest level of competence, have in effect been able to evade the spirit, if not the letter, of the regulations enforced by the KFTC.

It is important that the KFTC does not mix up competition aspects with policies for industrial restructuring (OECD, 2000c). The government has invested considerable effort and resources in creating market institutions, notably a reformed financial system and a new corporate governance framework. While serious structural issues remain to be tackled and require short-term remedies, market institutions will not be fully activated and are unlikely to play their intended roles as long as the government relies on direct control of the *chaebol*. For example, "non-market" transactions to assist affiliated companies have a direct impact on shareholders, who should use their powers under the new corporate governance framework to defend their position. Meanwhile, the KFTC should have a strengthened role in ensuring that markets are contestable by removing remaining barriers to entry and by establishing more open and competitive structures in vertically integrated industries. This will require a strengthening and redirecting of the KFTC's analytical competencies to enable it to serve as a true champion and guardian of contestable markets.

In addition, there is a need for careful follow-up on the implementation of corporate governance reform. Further measures should be taken to ensure accountability of business leaders to stockholders and financial institutions, and to loosen the uncontested hold of business leaders on corporate control. Attention should be paid to:

- *Corporate restructuring.* The procedures for dealing with firms in trouble should be improved by providing incentives for debtors and creditors to initiate re-organisation or composition proceedings prior to the closure of an enterprise and by giving a greater role to creditors in the governance of the debtor.

- *Shareholders' rights.* Rules expanding the use of cumulative voting could be established and companies should be required to schedule shareholder meetings in ways that would encourage, rather than limit, participation.

- *Corporate directors.* A clearer statement of the responsibilities and liabilities of corporate directors should be established; outside directors should be given greater responsibility.

- *Holding companies.* The restrictive conditions which, in practice, prevent the formation of holding companies should be eased in order to enhance transparency and accountability while promoting the sale of non-core companies.

Progress in these areas is vital for improving risk management and for moving away from a dependence on debt toward widespread equity financing, and for advancing the new forms of co-operation and alliances between firms which have become a significant instrument in knowledge and technology flows.

C. Improving conditions for small and medium-sized enterprises (SMEs)

In the OECD as a whole, SMEs make up over 95% of all enterprises (99% for Korea) and account for 60-70% of jobs (74% for Korea in 1997) (OECD, 1997; OECD, 2000d).[4] The share of SMEs in employment tends to be somewhat lower in manufacturing, ranging from 40 to 80% (69% for Korea). Table 6.3 presents some of the basic characteristics of industrial enterprises in Korea.

Korean SMEs fared particularly badly during the financial crisis. Average value added fell and some 20 000 small firms went bankrupt when the *chaebol* delayed payments to their SME suppliers. The problems confronting large segments of the Korean SME sector, however, are of a more long-term nature, go beyond the role of the *chaebol* and are related to the knowledge-based economy. While productivity differentials between large firms and SMEs have narrowed in some low-technology traditional industries – *e.g.* textiles, clothing and footwear, and food processing – they have increased in industries where technological competitiveness matters most – *e.g.* machinery, electronics, and transportation equipment. On average, value added per employee in SMEs relative to large firms fell from 55% in 1980 to 39% in 1997 (Woo, 1999). Meanwhile, the dearth of innovative SMEs has impeded the capacity of Korean industry to upgrade its production by entering into high-end niche markets (Ernst, 2000, p. 33).

Some important developments have started to alter the picture. In July 1996, KOSDAQ (a NASDAQ-like stock exchange) was established to facilitate equity funding of emerging businesses. In the wake of the crisis, special measures were also put in place, *e.g.* many more Korean SMEs obtained access to trade finance facilities and a discount window of the commercial banks. A separate program for SMEs was started with a focus on firms "in the middle" – *i.e.* those that were potentially viable but which needed support. Today, large banks are operating separate internal workout units for SMEs. With the *chaebol* in the process of reducing their debt exposure, the SME share in financing provided by the banking system improved markedly in 1999.

Table 6.3. **Basic characteristics of Korean manufacturing establishments, by size**

	1991	1993	1995	1997
Number of establishments in manufacturing industry				
Total	72 213	88 864	96 202	92 138
Percentage breakdown by size of enterprise				
SMEs	98.5	98.9	99	99.1
Large enterprises	1.5	1.1	1.0	0.9
Percentage breakdown by number of employees				
SMEs	63.5	68.9	68.9	69.3
Large enterprises	36.5	31.1	31.1	30.7
Percentage breakdown by value added				
SMEs	45.8	50.3	46.3	46.5
Large enterprises	54.2	49.7	53.7	53.5

Source: NSO, *Report on Mining and Manufacturing Survey.*

Following the easing of restrictions to entry and other accompanying structural and policy changes, there has been a surge of start-ups. About 3 000 start-up companies are established each month, compared to the exit of some 600 firms (Table 6.4). The entrepreneurs behind these start-ups are generally college graduates, a fair proportion of whom have strong engineering backgrounds. Women are also emerging as owners of venture businesses, with 2 546 women-owned start-ups in 1999, up from 1 447 in 1998. Virtualtek in the Internet business (recently listed on the KOSDAQ), and CardKorea provide examples.

Of the 3 000 start-ups per month, about 50 are new high-technology start-ups. From less than 100 in 1995, there were more than 3 500 high-technology start-ups in 1999. The prospects for technologically oriented, knowledge-based SMEs now appear exceptionally favorable. Not only does the government expect to support the growth of some 10 000 such start-ups though venture funds worth KRW 1 trillion (MOFE), but the changes in the *chaebol* and the economy at large bring new opportunities and particularly help to attract entrepreneurial resources and skilled labor, notably in knowledge-intensive service industries. High-technology start-ups have recently featured prominently on the list of venture enterprises maintained by the Small and Medium Business Administration.

At the same time, there has been an obvious overshooting in the evaluations of such enterprises, and an associated misallocation of resources and unwanted distribution effects (see section below on

Table 6.4.   **Trend of start-ups and bankruptcies, 1993-99**

Year	Seven metropolitan areas		Number of bankruptcies (national) (C)	A/B (Multiple)
	Start-up businesses (A)	Bankruptcies (B)		
1993	11 938	2 669	9 502	3.5
1995	17 245	4 559	13 992	2.9
1996	19 264	3 879	11 589	5.0
1997	21 057	6 132	17 168	3.4
1998	19 277	7 538	22 828	2.6
1999	30 701	2 479	6 718	12.4

Source: The Bank of Korea.

intangible assets). While the conditions for high-technology start-ups appear to have improved, the general conditions for start-ups and SMEs remain much less favorable. Although the SME sector has expanded since the mid-1980s, there has been little upgrading as regards technological capabilities and competitiveness.[5] The majority of SMEs remain engaged in low-technology industrial production. The sector remains thinly layered, with insufficient growth of firms capable of producing highly sophisticated parts, resulting in a weak foundation of firm-level technical training and technology development (Woo, 1999; Park, 1998).

Several factors interact to hamper the development of SMEs, including: product market competition and subcontractor systems; public procurement practices; formal and informal business practices, and access to skilled labor and technology.[6] For instance, there are serious shortages in technical competence, and accounting and marketing skills, and these shortfalls are being filled only slowly. Although the number of R&D labs in SMEs reportedly increased from about 500 in 1990 to some 3 000 in 1998, most SMEs remain mired in product adaptation rather than innovation. With respect to technology, there are significant gaps relative to internationally acceptable levels (KIET, 1998). At the same time, competitive pressures are set to increase, following from greater openness in the domestic market and because the *chaebol* are likely to become far more demanding customers.

Improved SME performance matters not only in its own right, but also because small firms are less likely to suffer "lock-in" with respect to existing plants, technologies and organisational structures, making them important for innovation and commercial experimentation with new technologies. At the same time, SME operations are typically characterized by high turbulence and churning, and the social benefits of their commercial experimentation tend to exceed the private ones. In the knowledge-based economy, policy intervention must be conducive to entrepreneurship and risk-taking, but should not shelter SMEs from change. Today, information and communication technologies open up new opportunities for combining the advantages of small scale with those from networking among SMEs (and/or between SMEs and larger firms, or between firms and other actors such as research institutes). Networks can also serve as an instrument to enable government policy to reach out more effectively. It is essential that networks are driven by the identification of market opportunities, however, and policies should be conducive to such "demand-led" approaches. The available experience suggests that successful network development requires a combination of measures facilitating the provision of venture capital, public procurement, technology diffusion, programs and incentives conducive to training, regulatory reform, etc. (UNIDO, 1995a; 1995b; 1999; Schmitz and Nadvi, 1999).

So far, the Korean Government has tried to balance market openness with selective support (Small and Medium Business Administration, 1999). A number of good programs have been introduced, aiming at, for example, the promotion of competitiveness through technological upgrading, deregulation and enforcement of competition law; the promotion of venture business through "business-angel" capital; assistance to entrepreneurs establishing new firms using their own technology; and the encouragement of technological innovation (as discussed in Chapter 5). In effect, however, much of the government support provided to SMEs to date has served to protect them from normal business pressures, leading to a dependence on government programs that has led to diminished competitiveness and innovation. The government needs to develop a comprehensive strategy aimed at encouraging improved competitiveness while scaling back direct intervention measures. In particular, it should replace protection by measures to help viable SMEs grow. It would be desirable to evaluate the cost efficiency and the economic outcomes of the myriad of SME programs currently on offer, partly as a means to ensure complementarity and avoid duplication. There is also a need for an effective mechanism for feeding the results of evaluations back into policy design (OECD, 1998).

In short, there is a case for:

- Consolidating present programs and increasing consistency among different policies, phasing out those that serve to shelter SMEs from healthy competition.

- Enhancing markets and programs to strengthen the diffusion of technology and skills.

- Improving the broader entrepreneurial climate through the creation of a social and economic environment – including government institutions – which is more friendly to small-business development.

D. Strengthening the contribution of foreign firms

Another feature of the Korean set-up until recently has been the limited presence of FDI, especially in the form of mergers and acquisitions, which has impeded foreign knowledge transfers. The ratio of FDI to gross fixed capital formation has averaged less than 1% over the past three decades (for China, this ratio has grown from virtually nil during the late 1970s, to close to 14% during 1992-96) (UNCTAD, 1998). The very high rate of output growth indicates that Korea has been able to compensate for this deficiency in other ways, *e.g.* through very effective application of imported technologies and large capital investments. However, there have been downsides. By relying on short-term bank loans rather than on long-term capital, Korea became susceptible to turbulent foreign exchange markets. In addition, the absence of foreign firms reduced competitive pressures, stifling one of the stimuli for innovative activity.

In recent years, Korea has taken major steps to liberalize its inward investment regime and reform bankruptcy regulations, etc. (see Chapter 2). Together with the decline in asset prices and a more welcoming attitude to foreign investors following the Asian crisis of 1997-98, this has led to a substantial increase in inward FDI (see Figure 2.1). There has also been a shift toward foreign investment in business services, especially securities firms, insurance companies and commercial banks, as part of the financial sector restructuring efforts. For 1999, foreign investment reached USD 15 billion, and the surge has continued into the first quarter of 2000 when it was up 36% on a year-on-year basis. EU companies have become the largest investors, taking over from the United States and Japan, and signaling a diversification of investment sources.

This development has helped to reinvigorate the economy by bringing in capital, technology and a package of soft core assets in accordance with global management standards. It has increased the potential for Korean actors to link into international networks and alliances. Increased competition and links to international systems of knowledge generation, together with growing FDI and other forms of collaboration with foreign multinationals, *e.g.* in the form of strategic alliances, will greatly influence the progress that can be made in knowledge-based activities. Foreign investments in companies have brought about changes in corporate culture, with traditional hierarchical management structures being increasingly discarded in favor of more transparent and efficient management. Mergers such as that of Hansol PCS with Bell Canada International have brought transparency to management and boosted efficiency. Doosan Seagram, and Oracle Korea appear to be creating flexible, ability-oriented and specialist organisations.[7] Foreign companies have also been aggressive in hiring Korean women professionals and this is having a demonstration effect on Korean companies. Samsung, for instance, is reportedly hiring a greater number of women professionals, partly as a result of the successful experience of foreign firms.

Nevertheless, a combination of formal and informal barriers continues to hamper the role that could be played by foreign firms in lifting the overall competitiveness of the economy, and has prevented them from providing a countermeasure to the power exercised by the *chaebol*. The Korean policy set-up retains features which limit the extent to which foreign investors are prepared to engage in extensive transfer of technology and skills (OECD, 1998). It is true that technology from abroad has played an important role in Korea's rapid development, but this applies primarily to technology embodied in imports of capital and goods, and licensing arrangements. Following the liberalization of licensing arrangements in 1978, royalty payments rose from an annual average of less than USD 100 million to USD 1 billion by 1990 and to almost USD 2 billion by 1995. Participation by Korean businesses in joint ventures with multinational corporations (MNCs) for the purpose of acquiring access to technology from foreign firms has been much more problematic. As Korean firms have become strong competitors in many fields, foreign companies have become uneasy about technology transfers.

Weak spots include the lack of high-skilled human resources and capable suppliers with technological and research know-how. In contrast with other countries that successfully attract foreign technology and know-how through inward investment – notably the United States, but also small countries such as Singapore and Ireland – Korea has severe barriers to imports of foreign expertise and talent: foreign managerial positions are few and mobility between science and research institutions and industry is seriously limited. Added to the small number of SMEs showing research and innovation potential, the inadequate pool of human resources and capable supporting firms runs the risk of hampering the country's ability to continually attract world-class MNCs.

The attraction of large inward FDI is not necessarily an indication of favorable economic prospects. Rather than the capital flows involved, the impacts on economic and social development crucially depend on how firms reorganize their operations and on the new skills and technologies made available through these changes in corporate behavior. Studies of foreign investment decisions point to inadequate protection of intellectual property rights and a weak climate of innovation, limiting the willingness of foreign investors to move beyond mere marketing activity and engage in production and sizable technology transfers (Mansfield, 1994; Maskus and Yang, 2000; OECD, 1998; KIET, 1998).[8] For example, in the biotechnology industry, Korean firms have mainly imitated foreign technologies in their production of antibiotics and vaccines. In the case of the software industry, the poor protection of IPR in Korea has hampered the development of the industry (KIET, 1998). Without improvements in these respects, even sizable inward investment flows might end up making only a modest contribution to the performance of the Korean economy, including to its advance into the knowledge-based industries and activities.

The Government of Korea should complement its extensive liberalization of foreign business activity with measures to strengthen the incentives for foreign firms to transfer technologies and skills. This agenda should include: the implementation of corporate governance reform; the strengthening of national innovation systems and intellectual property right protection; and the upgrading of human capital and measures to enhance the mobility of personnel, including dismantling the barriers to mobility between foreign and domestic firms. Improving the level of English-language knowledge and skills, already discussed in Chapter 3, would enable Koreans to work in FDI firms and gain higher positions in foreign-owned entities, and would enhance the learning effects to be gained from serving in such positions. In addition, the following measures should be considered:

- Improve the provision of information to potential or actual foreign investors, setting out the risks and opportunities associated with ventures in Korea ("Invest-in-Korea Services"), and particularly with respect to opportunities in knowledge-intensive areas.

- Introduce measures to catalyze a more effective domestic market for business services.

- Strengthen co-operation between domestic and foreign firms in research and development through the expansion of foreign participation in government-supported R&D projects.

- Carry out an examination of how other economies, including Ireland, Singapore and Chinese Taipei, have encouraged foreign firms to develop backward linkages in order to strengthen the technological capability and competitiveness of domestic suppliers.

E. Promotion of high-value-added services

Services represent a rapidly growing share of knowledge-based activities, although this development has taken place at a slower rate in Korea than in most other OECD countries (Chapter 1). A recent review of knowledge-based services[9] indicates that their share of business sector value added amounted to some 30.6% in 1996, compared to an OECD average of 41% (OECD, 1999d). Communications services (at 2.4%) were comparable to the OECD average, while financial, insurance and business services (at 19.5%) trailed many other OECD countries by 5 to 10 percentage points. The last years have seen a certain degree of catching up, with overall growth in knowledge-based services in Korea averaging 10.6% during 1987-96 (compared to 3.2% for the OECD area as a whole).

The knowledge-based economy is highly dependent on a competitive services sector. The more interlinked the provision of goods and services, the more technologically advanced the products and the greater the capacity to satisfy increasingly sophisticated consumer demand. Furthermore, as incomes rise, the demand for services tends to increase faster than the demand for goods. The production of services is less standardized and less capital-intensive (but typically not less knowledge-intensive) than the production of goods. Meanwhile, technical progress, in particular in ICTs, partly by facilitating storage and trade, is exerting a major impact on the functioning and organisation of many services. "Knowledge-based services" now account for a larger and more rapidly growing share of the economy than "knowledge-based manufacturing" in practically all OECD countries (OECD, 2000e).

Another set of factors influencing the role of services relates to industrial organisation. In many OECD countries, there has been a pronounced shift in the way that businesses organize service-related functions. To a growing extent, manufacturing firms tend to outsource what they perceive as their non-core functions – thereby fuelling an apparent growth in the service sector. The trend toward greater out-sourcing is typically driven by:

- *Competence*. The increasing sophistication of information, financial, computer, research and training needs by business, and the rapid evolution of new techniques and products in these fields has made it difficult for firms to maintain competitive competence in these areas. To do so would require the accumulation and maintenance of a knowledge base in diverse disciplines which, in most instances, firms would be hard-pressed to justify.

- *Cost and efficiency*. Firms specializing in support functions are often able to provide their services at lower cost, while offering a w7ider choice of innovative products, reflecting the positive effects of competition (in-house services were likely to be shielded from such competition, a condition which lowered the need to maximize efficiency and to innovate).

- *Specialization*. The trend in recent years has been toward consolidation and concentration on core competencies, providing increased opportunities for independent suppliers of goods and services.

Outsourcing by large and small firms alike is set to grow in Korea in tandem with the emergence of innovative, knowledge-based service providers. However, in Korea and, to a lesser extent, Japan, this trend has so far been weaker than in most other OECD countries; companies have preferred to retain control of many service-related functions in house. Obstacles to outsourcing in Korea include the emphasis on vertical hierarchical structures, and barriers to the mobility of workers. In addition, as noted in Chapter 2, a number of regulations continue to constrain the service sector's ability to innovate, adjust and grow. There is no panacea for stimulating greater competitiveness and higher growth in business services. However, several areas require attention:

- Continuing efforts to remove regulatory barriers to market entry and exit.

- Initiating creation and expansion of new service markets by promoting outsourcing of services in the public sector.

- Reforming financial support to innovation and government-supported research programs to make them more applicable to the service sector. Current incentives are primarily geared to the needs and opportunities of manufacturing operations.

- Reviewing education and training policies to support the development of the human resource skills needed to support knowledge-based industries, most of which are service-oriented.

- Exploring ways to adapt intellectual property rights to more effectively meet the needs of service industries.

- Improving data on service industries; this will benefit service providers, users and investors alike, while providing the basic information required by governments to formulate policy.

- Continuing work in the WTO-GATS framework to identify effective strategies for liberalizing trade in services on a multilateral basis.

- Fostering a better functioning market for knowledge-based consultancy services, without resorting to public consultancy services as these can impede the development of private consultancy services.

- Improving conditions for investment in intangible assets (see below).

F.   Promoting investment in intangible assets

Traditionally, both the private sector and the government have focused their attention on investment in physical assets, such as machinery and equipment. In the KBE, however, the source of value has shifted from physical content to knowledge content. The development of knowledge-based activities is critically dependent on conditions conducive to investment in so-called intangible assets (*e.g.* innovation, worker skills, patents, brand names). The weak state of services in Korea, for instance, is related to a bias in favor of investment in plant and equipment relative to investment in intangible assets.

Today, corporate investment in intangible assets is growing faster than tangible investment throughout the OECD. Investment in knowledge (defined as R&D, software, public spending on education) now represents an estimated 8% of OECD-wide GDP, a figure similar to investment in physical equipment. Investment in intangibles ranges from between 9-10% of GDP in the Nordic countries and France to 6-7% in Italy and Japan.[10] The OECD average exceeds 10% when private spending on education and training is included (OECD, 1999*a*). Yet, this is only a partial picture. For a list of possible components of investment in intangible assets, see Box 6.1.

Problems arise partly because investments in intangibles tend to assume characteristics of public goods, making it difficult for private investors to capture the full benefits of their investments, and leading to under-investment. Such market failure arguments have long been the rationale for government intervention in areas such as science, research, education and training. In addition, the inability of enterprises to effectively report to capital markets on the extent, importance and utilization of intangible assets reduces their incentives to improve internal management systems (OECD, 1998).

Venture capital is important for enabling investment in intangible assets, especially in new, explorative firms and industries. Risk-taking is a key element in venture capital, and excessive government guarantees regarding the financing of intangibles can counteract the innovativeness that they aim to encourage. The problems with measuring and disclosing intangible assets are particularly severe in a debt-oriented economic system such as Korea, since intangibles cannot generally be taken as collateral for bank loans. This is especially troublesome for new firms without a track record, and in services where tangible assets tend to be limited. The prevailing situation in Korea in this regard may appear paradoxical, since some new enterprises within these categories have recently found themselves drowning in equity funding. However, where systematic information on intangible assets is hard to come by, it is particularly difficult for the market to evaluate which new firms are viable and which are not.[11] Although a similar situation is observable in other OECD countries, there are special complications in Korea. These are partly rooted in "cultural factors", such as the lack of preparedness to adopt new ideas. They are also a result of the traditional weakness in providing information about business operations, which has brought about a habit of reliance on inadequate information in investment decisions, providing little shelter from overly high expectations in new growth areas. Whereas any excesses in that respect should be expected to correct themselves sooner or later – although inevitably resulting in some painful lessons and unwanted distribution effects – the current situation should lead to a strengthening of the demand for better transparency in business operations.

For the majority of companies which have been unable to turn limitations in information management into an advantage, the prevailing situation is hampering service sector development and entry by newcomers. In response, the government has developed a system of valuing technology assets for the purpose of strengthening access to finance particularly for new entrepreneurs. A financing institution has been set up to provide technology warrants or guarantees for banks that use these warrants as collateral for loans. Loan conditions have been eased for SMEs with weak credit ratings and little collateral, with credit guarantees for SMEs which are unable to provide collateral. It is not only start-up and survival that matters, however. The effectiveness of public funding for high-technology start-ups needs to be better assessed and benchmarked against the situation in other countries.

There is also a need for initiatives to increase the understanding of intangible assets. Most important among these is to promote initiatives by firms themselves. While individual firms in several OECD countries have recently taken steps to improve their measurement and reporting of intangibles, this tends to be anecdotal and is performed in a way that allows for little systematic comparison and analy-

---

Box 6.1. **Components of investment in intangible assets**

1. **Computer-related**

    - Software
    - Large database
    - Other computer service

2. **Production and technology**

    - R&D
    - Design and engineering
    - New quality control systems
    - Patents and licenses
    - Know-how

3. **Human resources**

    - Organized training
    - Learning by doing
    - Activities which improve the health and motivation of the workforce (including labor relations, physical check-ups, and other sport and fitness programs)

4. **Organization of the firm**

    - New methods of organisation of the firm as a whole
    - Establishment of networks
    - New working methods

5. **External: marketing and sales**

    - Market research
    - Advertising
    - Brands
    - Firm name and logo
    - Lists of customers, subscribers and potential customers
    - Product certification, quality certificates

6. **Industry-specific**

    - Quotas (which may not be traded and thus lack a market price), *e.g.* in mineral exploration
    - Entertainment, literary and artistic originals

Source: Young (1998).

---

sis. The challenge for the Korean Government is to stimulate the adoption of comparable and relevant practices without stifling creativity and differentiation in reporting – crucial because of the inherently idiosyncratic nature of intangible assets.

The valuation of intangible assets is influenced by attitudes. The government can exert a certain influence in this respect as well. The following are a few concrete initiatives that may be considered for building a more hospitable policy set-up for investment in intangibles in Korea:

- Publicizing and broadly disseminating cases of successful development of intangible assets by Korean citizens and companies, such as patenting in the semiconductor area; the successful development of Korean software, like the Hangol word-processing package and Korean game software; and cultural and literary achievements.

- Broadening the support for innovative efforts away from traditional R&D support.
- Examining options for introducing tax incentives to stimulate training (*e.g.* exempting profit tax on profits used for paying for personnel training).
- Reviewing and correcting government policy bias in favor of physical investment.
- Promoting research and evaluation to document successful cases (best practice) of foreign-Korean co-operation, and developing the means to diffuse such information effectively.
- Improving the enforcement of intellectual property right protection.

G. Conclusion

The dominance of the *chaebol*, the fragile SME sector, barriers to contributions by foreign investors, the weaknesses in knowledge-based services, and inadequate conditions for investment in intangible assets combine to weaken the mechanisms for the development of knowledge-based activities and industries in Korea. This chapter has emphasized that the Korean Government should continue to press for efficient resource allocation and higher profitability in the *chaebol*, but should do so primarily through the financial channel and through corporate governance reform. Meanwhile, competition policy should be clearly separated from structural, or interventionist, policy. Conditions for start-up and growth of SMEs should be improved through a consolidation of existing support programs backed by more critical evaluations and an increased emphasis on the diffusion of technology and the upgrading of workforce skills. Furthermore, the government should follow up on the liberalization of inward FDI with measures to induce foreign firms to transfer more skills and technologies to Korea, including improved intellectual property rights and the removal of barriers to labor mobility. It should dismount the remaining barriers to service-sector development, *e.g.* through continued regulatory reform, and develop a comprehensive strategy for improving the conditions for investment in intangible assets.

Appendix

Most Important Bottlenecks in Production and Technological Improvement

Ranked from 1 = "no problem" to 5 = "major problem"

	Average	Standard deviation	Number of responses
a. Lack of high-quality local suppliers	2.96	1.22	809
b. Lack of machinery and equipment supplier and service companies	2.66	1.12	807
c. Infrastructure service (water, electricity, telecommunications, transportation)	2.18	1.03	804
d. Quality and supply of technicians and high-skilled personnel	3.23	1.15	821
e. Quality and supply of production workers	3.09	1.11	817
f. Labor costs	3.29	1.07	822
g. Minimum wage laws	2.28	1.06	805
h. Restrictions on hiring or firing of workers	2.56	1.12	805
i. Labor unrest/strikes	2.24	1.22	806
j. Access to finance for working capital	3.37	1.26	829
k. High interest rates make loans too costly	3.84	1.19	828
l. Customs administration	3.09	1.21	792
m. Red tape/bureaucracy	3.58	1.15	818
n. Corruption	3.58	1.15	818
o. Protection of intellectual property rights	3.09	1.26	799
p. Import tariffs and quotas on inputs, raw material and equipment	2.69	1.15	792
q. System of import duty exemption/reduction	2.88	1.19	788

1. The survey was distributed to 2 500 enterprises, of which 863 responded. It was initiated in November 1998 and completed in February 1999. SMEs (20-300 employees) made up approximately 85% of the sample; about 13% were firms with more than 10% foreign ownership. The differences in the average scores between the large and small sub-segments was not very significant – except for corruption which was a bigger problem for SMEs.

Source: World Bank supported survey of Korean enterprises.

Notes

1. Total business sector is defined as ISIC Rev. 2, Divisions 1 to 9.

2. *High-technology manufacturing* includes computers, aerospace, pharmaceuticals, and communications equipment. *Information and communication manufacturing* includes office and computing equipment, radio, TV and communications equipment; *services* includes communications services. *Knowledge-based industries* also includes finance, insurance and business services.

3. The table seeks to illustrate the structural strengths and weaknesses of Korean knowledge-based industry. The selected industries belong to the high-technology category, as defined by the OECD, and are viewed either as strongholds of Korean industry or as "future industries".

4. Definitional differences complicate comparisons. While SMEs are usually defined to include firms with less than 500 employees, a number of countries use a lower cut-off point of 250 (300 in the case of Korea).

5. A November 1999 KOTRA survey of 200 buyers indicated that the principal problems encountered by SMEs, in order of importance, were: *i*) price competitiveness; *ii*) observance of shipment deadlines; *iii*) after sales service; and *iv*) complaints from European and US buyers that Korean SMEs lagged in new product development. Other problems identified included late invoice payments; lack of an aggressive long-term marketing strategy and quality control (the Japanese complained of defective products) (Sim, 1999).

6. Based on a survey of five industries representing the largest sectors in Korea in terms of value added and exports. Given that the survey was undertaken in the middle of the financial crisis, it is not surprising that access to credit and high interest rates were a major problem. Red tape and corruption are still important impediments for Korean firms and this, together with the relatively high average cost of protection of intellectual property rights, substantiates the impression that problems remain in the broader institutional regime. The high rating for difficulties relating to the quality and supply of technicians and high-skilled personnel highlights the challenges outlined in Chapter 4. Annex 1 includes a table on the most important bottlenecks facing Korea in terms of improving production and technological know-how .

7. A good knowledge of English is a means of survival for Korean staff in foreign-invested companies. LG telecom, together with the United Kingdom's BT, sponsors six months of language training for its employees in the United Kingdom. Other companies are changing the corporate landscape: Volvo Korea and Korea Fuji Xerox operate in-house language classes (Business Korea, 1999).

8. The impact is strongest in activities such as drugs, chemicals, machinery and equipment, and electrical equipment. However, it appears that foreign investors perceive the intellectual property right regime as an important indication of the attitude of the host country regime toward foreign investors, which is important for their general willingness to make long-term commitments of the sort required for setting up local R&D facilities, for example.

9. For analytical purposes, knowledge-based services are defined to include communications services, finance, insurance and other business services, and community, social and personal services.

10. Korea is not included in this ranking, due to a lack of comparable data.

11. On 10 March 2000, for instance, the *Wall Street Journal* reported how mere advertising on the Internet enabled inexperienced Korean entrepreneurs to attract millions of dollars. On 3 April, *Maeil Business News* reported that 27 KOSDAQ companies were temporarily suspended for failing to meet trading rules.

References

Ernst, Dieter (2000),
"Catching-up and Post-Crisis Industrial Upgrading: Searching for Sources of Growth in Korea's Electronics Industry" in F. Deyo, R.F. Doner, and E. Hershberg (eds.), *Economic Governance and Flexible Production in East Asia*, forthcoming.

Korea Institute for Industrial Economics and Trade (1998),
"Profile of Korean Industry", Seoul: KIET.

Mansfield, Edwin (1994),
"Intellectual Property Protection, Foreign Direct Investment, and Technology Transfer", *International Finance Corporation Discussion Paper*, No. 19, *http://www.ifc.org/economics/pubs/dp19/dp19.pdf*.

Maskus, Keith E., and Guifang Yang (2000),
"Intellectual Property Rights, Foreign Direct Investment and Competition Issues in Developing Countries", *International Journal of Technology Management*, Vol. 19, Nos. 1/2, pp. 22-35.

OECD (1997),
"Small Business, Job Creation and Growth: Facts, Obstacles and Best Practices", free brochure, Paris: OECD.

OECD (1998),
Technology, Productivity and Job Creation: Best Policy Practices, Paris: OECD.

OECD (1999a),
OECD *Science, Technology and Industry Scoreboard 1999: Benchmarking Knowledge-based Economies*, Paris: OECD.

OECD (1999b),
Asia and the Global Crisis: The Industrial Dimension, Paris: OECD.

OECD (1999c),
OECD *Economic Surveys 1999: Korea*, Paris: OECD.

OECD (1999d),
Strategic Business Services, Paris: OECD.

OECD (2000a),
"Knowledge-based Industries in Asia", Paris: OECD.

OECD (2000b),
OECD *Economic Surveys 2000: Korea*, Paris: OECD.

OECD (2000c),
Regulatory Reform: Korea, Paris: OECD.

OECD (2000d),
OECD *Small and Medium Enterprise Outlook*, Paris: OECD.

OECD (2000e) (forthcoming),
"The Service Economy", free brochure, Paris: OECD.

Park, Joon Kyung (1998),
"Creating an Extra-firm Infrastructure of Institutions for Small and Medium-sized Businesses", in Lee Jay Cho and Yoon Hyung Kim (eds.), *Korea's Choices in Emerging Global Competition and Cooperation*, Seoul: Korea Development Institute.

Schmitz, Hubert, and Khalid Nadvi (1999),
"Clustering and Industrialisation: Introduction", *World Development*, 27(9), pp. 1503-1514.

Sim, S.-T. (1999),
"Number of Start-up Companies Expected to Surpass 30 000 in 1999", *Korean Herald*, 15 December.

Small and Medium Business Administration (Korea) (1999),
Small and Medium Business Administration 1999.

United Nations Conference on Trade and Development (UNCTAD) (1998),
World Investment Report, Geneva: UNCTAD.

United Nations Industrial Development Organisation (UNIDO) (1995*a*),
"Principles of Promoting Clusters and Networks of Small and Medium Enterprises", *http://www.unido.org/doc/Publications.htmls*.

United Nations Industrial Development Organisation (UNIDO) (1995*b*),
"Industrial Clusters and Networks: Case studies of SME Growth and Innovation", *http://www.unido.org/doc/Publications.htmls*.

United Nations Industrial Development Organisation (UNIDO) (1999),
"SME Cluster and Network Development in Developing Countries", *http://www.unido.org/doc/Publications.htmls*.

Woo, Cheonsik (1999),
"Inbound FDI and Industrial Upgrading of Korea: Prospect and Challenges", Seoul: Korea Development Institute.

Young, A. (1998),
"Towards an Interim Statistical Framework – Selecting the Core Components of Intangible Investment", *http://www.oecd.org/dsti/sti/industry/indcomp/prod/paper3.pdf*, Paris: OECD.

Chapter 7

Implementing Korea's Strategy for a Knowledge-based Economy

A. Introduction

Since early 1999, the government has been developing its vision for making Korea an advanced knowledge-based economy, as announced by President Kim in January 2000. The Ministry of Finance and Economy (MOFE), in conjunction with 13 think tanks led by the Korea Development Institute (KDI), prepared a series of background papers which were discussed at a public hearing in October 1999. This was followed by further work on the overall strategy, and presented to the subsequently created National Economic Advisory Council (NEAC) and the President.

In January 2000, following the Presidential announcement, the NEAC unveiled a three-year master plan focusing on five key themes: *i)* developing the national information infrastructure; *ii)* improving national science and technology innovation capabilities; *iii)* developing new knowledge industries and digitizing older industries; *iv)* developing the human resource system to respond to the knowledge-based economy; and *v)* addressing the "digital and knowledge divide". In the first quarter of 2000, five task forces (of 14-18 members each) from the responsible ministries, agencies and research institutes were created around these broad strategies and began preparing detailed action plans for each of these themes. Another public hearing was held in April 2000 on the draft Action Plan. The details were then approved by the Cabinet and put into effect in May 2000 (KIET, 2000).

The Economic Policy Co-ordinating Committee (EPCC) in the Prime Minister's Office is expected to be made responsible for the overall implementation of the Action Plan, including the co-ordination of policies and measures and the updating of procedures. Implementation is expected to be carried out through the same type of multi-agency framework as is found in other countries, by all the respective agencies participating in the task forces. In early May 2000, a co-ordinating Task Force was established, made up of MOFE and the Ministry of Planning and Budget (MPB), some civilian experts and the heads of the five above-mentioned task forces. The Director General of MOFE's Economic Policy Bureau will preside over the co-ordination of the task force, and will be responsible for all practical matters (Woo, 2000).

These are important steps. We elaborate below on two issues which, in our view, will strongly influence the extent to which this action plan can be successfully implemented. The first relates to the process of consultation in Korea which, given the social and political changes taking place, needs to be broadened and deepened. The wider consultations with society which take place in many other OECD countries are not yet the norm in Korea's mode of government. The second issue concerns the practical workings of the implementation plan, which need to be further refined.

B. Broadening and deepening the consultations

1. *Overcoming resistance and changing mind-sets*

The very nature of the knowledge-based economy implies distributed power, with far-reaching consequences for society as a whole. It is therefore important that the agenda for reform is also owned by the private sector and civil society to ensure that they not only understand the key trends and forces affecting them and therefore the need for change; but also what the implementation of that agenda will require of individuals, firms, organisations and the government. In this regard, the chosen implementation strategy should closely fit the country's institutional realities. This will involve policy reforms, government initiatives, institutional re-organisation, and actions by firms and social groups.

Some elements of the reform package can be implemented by modifying regulations or procedures. Others may require changes in legislation (*e.g.* the regulatory framework for telecommunications). Such legislative reforms may require lobbying in order to convince legislators of the need for change. There may also be resistance to many of the proposed reforms, some of which may come from interest groups who stand to lose power or authority; or it may stem from a misunderstanding of the rationale for change. Other forms of opposition may be based in tradition or entrenched beliefs (*e.g.* that the educational system should be supply-oriented or that knowledge is a free good).

Developing and effectively implementing the strategy will entail challenges in the short, medium and long terms; and these should not be underestimated. Based on the experience of other countries that have crafted and implemented such broad strategies (Canada, Finland and Ireland), it is important to emphasize that the development of the strategy must be undertaken in consultation with the private sector and civil society.[1] This is particularly relevant in Korea given the current social and political change that is taking place and in light of the country's vision for the 21st century. The public hearings[2] and consultations that have taken place with a number of think tanks and private sector bodies on national policies and reforms, are an encouraging sign. However, building consensus and getting stakeholders to "buy" the proposed measures will require greater efforts in terms of dissemination, explanation and consultation with a wider range of experts and other actors in society. This will require highlighting the broader societal and public interest aspects of the proposed changes as being necessary to help Korea continue to improve the income and welfare of its citizens in an increasingly complex and competitive world, where more effective use of knowledge is critical to enable the country to remain competitive as well as to deal with the constant restructuring caused by increasingly rapid advances in knowledge.

In addition to stakeholder involvement, a key element in any reform is opening up established attitudes to critical reflection. This can only be accomplished through dialogue and persuasion. As the Finnish and Irish experiences have shown, it is not possible to steer the development of the KBE in a centralized manner. Some examples of attitudes requiring re-examination include: the fear of taking entrepreneurial risk; the preference for university vs. business jobs among the highly qualified; the low self-esteem of vocational degree holders, etc. Some of these attitudes may change slowly through the impact of a greater openness to market forces but a conscious effort to promote change can be made through wider dissemination of information (*e.g.* on returns to different types of jobs) and in collaboration with the media. Again, as Finland has demonstrated (see Box 7.1), a broader consultation of sectoral interests (education, science and technology, ICT access, women, labor, etc.) can go a long way to ensuring an enduring vision and plan of action.

Box 7.1. Finland's broad consultations

Broad consultations were conducted by Finland's Steering Group for Strategy Updating (SGSU) with the National Council for Information Society (NCIS) and the Information Society Forum (ISF). Relevant institutions were invited to take part in the discussions, background reports were produced and experts interviewed, several brainstorming events and seminars were held, relevant ideas were received through 5 000 e-mails, and the first draft proposal was circulated to 150 experts for comments and feedback. The project had a full-time secretariat, with a project manager. All these inputs were used by the Steering Group in formulating Finland's implementation strategy.

2. Korea's educational reforms: improving the consultation process

To illustrate the consultation process, we use the example of reforms in the education sector, and human resource management more broadly. Many of the proposed changes have been formulated by Korean think tanks and concerned groups. However, much of the resistance to reform is likely to come from the education agencies and from teachers, as both groups stand to lose some power and control. An important task is therefore to convince them of the benefits of the proposed reforms. Likewise, without the active support of teachers, it will not be possible to implement the move to a student-centered education system. Therefore, it will be essential to raise awareness and to invest in retraining teachers as part of the reform process.

A *complex process*. At the same time, the education reform process is complex and will take time; there is no instant solution guaranteeing quick results. Policies alone are not sufficient for change – what counts is their implementation. Global experience and research in education reform show that factors such as a shared vision, clear goals, strong leadership, broad consensus building, the ability to make adjustments throughout the process, measurable indicators and the availability of a support infrastructure are critical to ensuring that the reforms are implemented and achieve the desired results. In short, the systemic reform in education required in Korea will be a long-term process. The challenge is how to maintain the good practices within the existing system, such as relatively high average quality and low dispersion of student performance, while trying to create a new system that will serve the country's human resource needs in the global economy.

Lessons from earlier reforms. Korea has been engaged in a series of education reforms since the early 1990s. Lessons from these reforms suggest that the current reform process requires substantial adjustment. The complexity of the agenda and the need for support, co-ordination and integration of such reforms in Korean society at large, will require that the government build up greater public awareness of what is at stake. It may also be necessary to reconfigure some of the government ministries, for example, labor, education, and science and technology, that are involved in education and training. The Korean Government should adopt a more integrated approach to reform in education and human resource management, and appoint a unified national governing body to oversee these reforms and their implementation. This body should comprise representatives from the key stakeholder groups, and should not be perceived as being controlled or steered by the government. As a start, it should review the PCER-proposed reform programs and their implementation in the light of a clearly articulated conceptual framework involving education, human development, the changes in the global economy, and the requirements of the knowledge-based economy. Special attention should be paid to the vision, goals, appropriate timeframe and content of the reform, as well as to the lessons learned from earlier reform efforts, in order to make the necessary policy adjustments.[3]

Negotiations with stakeholders. Based on the outcome of such a review, a process of negotiation with key stakeholders – from industry, local government, teachers' unions, school administration associations, parents and representatives from civil society – on reform-related matters should take place in order to share the new vision and build the necessary consensus for reform. In particular, parties opposed to the reform should be included in this process. Constructive inputs from all involved stakeholders should be taken into account in policy adjustments in order to make the reform policies more legitimate and ensure broad ownership. This is not an easy process. However, experiences from reforms around the world (from Australia, the Netherlands, New Zealand and Singapore) indicate that managing this process will be critical in determining the ultimate success of the reform of the Korean system for education and human resource management.

Awareness campaigns. Once consensus has been built, the government should launch a well-designed and massive campaign to build public awareness of the need for such a profound education reform for the future development of Korea – what needs to be changed, how the proposed changes will take place, who will be involved and what are the expected outcomes. Such a campaign will require more than simply publishing brochures, posters and advertisements on TV or other media. The campaign should aim to "sell" the reforms to the public. The communication strategy for the recent public campaign for the National Literacy and Numeracy Strategy in the United Kingdom provides a good example in this regard (Cole, 1999).

Lifelong learning. To successfully achieve educational reform, the scope of the exercise should cover not only the formal education sector but also lifelong learning. The crucial role played by lifelong learning should be emphasized because it provides a tool for addressing the learning needs of individuals throughout their life time, and complements the weaknesses of the formal education system. Lifelong learning involves both formal and informal learning. The importance of informal learning, especially at the enterprise level, is on the rise. New forms of learning are becoming available, such as through the Internet, increased use of TV-based instructions, distance learning, virtual universities, etc. These trends will intensify in knowledge-based economies and call for the integration of the current formal, vocational, adult and distance education systems with instruments such as the Education Credit Bank system.[4]

To open the Korean education system to the world, an international expert panel should be created to bring in knowledge on global best practice and reform experiences, and to provide objective and comparative studies to Korean educators and policy makers. Internally, the curriculum on English teaching and learning should be strengthened and exchange programs with international education institutions should be expanded. In addition, women's education in scientific and technical areas, as well as in overall economic activity, should be fostered. These measures, plus the necessary resources, will provide a sound infrastructure for the reform effort. With this foundation and overarching framework in place, educational institutions will be able to reconfigure and re-structure themselves within the appropriate regulatory framework to make the relevant changes in crucial matters such as the curriculum, teaching practice, instruction materials and administration, with appropriate incentives and the involvement of teachers, parents, students and the community at large.

C.   Working out the implementation details

Implementing major reforms and actions in an interlinked manner as proposed in this report, requires defining specific roles and responsibilities, co-ordinating implementation efforts and setting monitorable goals through action plans.[5] We have already noted the rapid progress made to implement the Action Plan. However, three critical questions need to be addressed in order to facilitate successful implementation: *i)* Who will co-ordinate, and participate in, the implementation of the plan? *ii)* How will the needed resources be allocated? *iii)* How will the plan be monitored, evaluated and adjusted?

1.   *Who will co-ordinate the strategy?*

In the case of government initiatives, it is natural that government agencies be called upon to implement and co-ordinate activities using existing mechanisms within each ministry or through Councils (*e.g.* the NSTC), and the Ministry of Planning and Budget (MPB) on the allocation of resources. For the private and voluntary sectors, the issue would be: Which organisation or organisations can best represent developments in their respective sectors?

An important issue here is the role of government and that of other actors. While the government has a critical leadership role in the implementation of the strategy, it needs the participation of other actors. Canada, Finland and Ireland used the shared responsibility principle as an overarching theme in their strategies. Co-operation between sectors and administrative fields and within the international context was highlighted. Public sector, industry and other organisations were invited to initiate, participate in the decision-making process and commit themselves to promoting the defined objectives. The stated principles should also be applied to decision making concerning the allocation of resources, reflected in the strategies and decisions of the different sectors and parties, in the ongoing projects and in the preparation of "spearhead" or "flagship" projects.[6] In addition, these strategies should focus on promoting the functioning of markets where possible through de-regulation and competition, while building up modern regulatory oversight.

2. How to allocate the needed resources?

Implementation of the strategies should also focus on whether the infrastructures to implement reforms and measures are in place, especially in terms of staff and resources. In the Irish case, the resources needed to ensure successful implementation were assessed by the concerned departments and agencies in consultation with the Prime Minister's Department, the Ministry of Finance and the Implementation Group in accordance with normal financial planning (Second report of the Irish Interdepartmental Implementation Group) (see Box 7.2).

Box 7.2. **Implementation and financing: The Irish case**

In the case of Ireland, at the time of the adoption of the Action Plan in January 1999, the Interdepartmental Implementation Group was directed by the Irish Government to assess the resources needed to implement the plan. This was the primary focus of the group for the first few months. In its second review in July 1999, the Group concluded that substantial additional staff and funding needed to be made available if rapid progress was to be made and the objectives of the Action Plan met. The resources needed to implement the Plan are very substantial and in most cases an intense period of up to three years is envisaged. An evaluation team was established as a fast-track mechanism to approve projects for funding. A policy development team was established to co-ordinate the implementation of the Action Plan. The Department of Finance approved a staffing envelope to facilitate the implementation of projects. A high-level cross-departmental Implementation Group was set up to co-ordinate and drive the process with a set of guidelines during project preparation and project approval and implementation.

Co-ordination of efforts is equally fundamental as there are many overlapping initiatives by related agencies in Korea that address a particular objective (e.g. developing science and technology, digital divide, development of venture firms, SME development).[7] The Finnish strategy explicitly aims to overcome overlapping mandates and promote synergy among projects with the aim of reducing costs. It has created development networks between existing and new projects in order to enhance knowledge and information transfer and for the congruence of services being developed to meet the desired needs, and to evaluate measures against the broader goals of the country's vision.[8]

Another fundamental issue for the successful implementation of the overall development strategy for the transition to a knowledge-based economy is to secure sufficient financial resources and to allocate them efficiently among the various elements of the strategy. Some of these elements will require reallocating investments or changing approaches rather than higher expenditures. Others, however, will call for increased resources. Any additional expenditures must be decided upon within the overall taxation and expenditure regime, with appropriate consideration being given to the effects on incentives, efficiency, equity and macroeconomic stability.

To ensure efficient allocation of financial resources, adjustments should be made to the budget process.[9] Recently, new measures have been taken to increase the efficiency of government expenditure. These include: the medium-term fiscal plan, a preliminary appraisal system for major social overhead capital (SOC) projects, attraction of private funds for SOC investments, etc. Continuous monitoring and feedback is crucial to the implementation process. For example, the multi-year approach adopted by the MPB goes some way to facilitating the necessary monitoring and adjustment in a co-ordinated and participatory manner (see Box 7.3).

Box 7.3.   **Korea's medium-term fiscal plan**

In order to overcome the limitation of annual budgeting, Korea's Ministry of Planning and Budget adopted a multi-annual budgeting process in 1998. This is the medium-term fiscal plan which lays out indicative targets for fiscal investment, sources of revenue, budget balance, etc. The plan is revised each year and adjusted to accommodate the rapidly changing environment. In the planning process, the MPB assesses competing budget demands among agencies and projects by evaluating the effectiveness of long-term programs, policies and procedures, thus improving the efficiency and flexibility of the annual fiscal policy. The plan has also contributed to enhancing the soundness of Korean public finance by co-ordinating national policies requiring large government expenditures.

Source: Ministry of Planning and Budget (1999).

---

3.   *How to monitor, evaluate and adjust the plan?*

The dynamism of the knowledge and information revolution and the rise of the global economy, emphasize the importance of setting up a monitoring and evaluation system for each of the broad policy goals and strategies in the functional areas, and creating procedures for adjusting plans and actions in accordance with the findings of the impact evaluations. In some cases, the broad goals may have to be refined; in others, policies and policy instruments may need adjustment. In yet others, better use of resources or co-ordination efforts might be required.

Who will do the monitoring and evaluation? Monitoring and review is expected to be carried out by EPCC, NEAC, NSTC, the Informatization Promotion Committee, and the Informatization Strategy Committee. The five task forces are expected to support the workings of the EPCC during the implementation phase. It will be important to consider how to incorporate the views of key stakeholders, including the private sector and civil society.

Evaluation will be needed at the project, intermediate and the broader macro level. It is important to obtain a sector-specific industry perspective as well as the regional perspective. It is essential for evaluations to go beyond narrow efficiency aspects to address economic impacts, and to do so at a level which enables economy-wide outcomes and linkages between measures in different areas to be addressed (*i.e.* avoiding a piecemeal approach).[10] It is important to evaluate productivity- and innovation-based competition, transformation of SMEs into knowledge-based entities, in addition to measuring numbers of start-ups, evaluation of "flagship projects" (*e.g.* e-government, the national information infrastructure). Program evaluation should cast light on the types of activities that are more or less successful and how programs can be better designed and managed. Also necessary would be the evaluation of education reforms and the digital divide. It would be useful for the MPB to be included in monitoring and for a program audit and evaluation to be conducted. In this regard, use of external experts would bring best practices found elsewhere into Korean practice.

What will constitute the monitoring indicators? How will the relevant statistics be obtained? This implies that the National Statistical Office should participate in the project in order to organize data specific to the KBE initiatives (and should later publish them from a holistic perspective). Other data inputs might also be useful, such as project and program surveys and assessments (*e.g.* computers in schools, lifelong learning indicators), monitoring of overall goals and comparing them with G7 countries (*e.g.* in educational standards, performance in science and technology, etc.). Other indicators might also be necessary, *e.g.* on the impacts of programs and instruments on the disabled, on the creation of local knowledge communities, etc. The private sector might be able to provide indicators on knowledge management by firms.

Box 7.4. Finland's rolling strategy

Process in the development of Finland's "information society" is a decision-making process with the support of different actors in society. The process seeks to identify and anticipate opportunities and threats as the country develops its information society. It is based on the belief that it is not possible to steer the development of the information society in a centralized manner given the rapid environmental changes. The strategy process focuses on *monitoring changes* in the environment and *adapting* the strategy accordingly. It uses extensive information society *statistics* and, with the help of a wide range of experts, reviews and revises the priorities. It assesses the social and societal effects of ICT and the need for new procedures and services as an important element of the review process. It also *disseminates information* concerning the monitoring and achievements of objectives of programs and projects on a short-term basis, complemented by a more comprehensive annual review. This review is carried out by SITRA, the Finnish National Fund for Research and Development, which reports to Parliament on an annual basis.

Source: SITRA (1998).

What is involved in continuous adjustment? Given rapid changes and new emerging issues, some anticipation of changes (*e.g.* technological forecasting; broader foresight studies)[11] and how they will affect the chosen strategy should be undertaken. It will also be important to work out how the findings from the monitoring and evaluation can usefully feed back into the implementation plans of the different actors. Again, Finland provides a possible approach in this area (see Box 7.4).

These are preliminary recommendations and should be treated as input to an ongoing process. This report addresses key aspects of a very dynamic system, identified by two international institutions which have the advantage of a broader perspective of cross-country experiences but which are not necessarily familiar with all the specifics of the Korean context. It should thus be interpreted as representing one contribution to a process which has to be locally owned – not only by the Korean Government but, more importantly, by the Korean people. We have attempted to provide a broad framework against which to highlight the relevant issues as we see them. Given Korea's own strong capacities, the country must set up its own reform process, drawing on global experience and adapting it to the specificities of the Korean situation.

Notes

1. Although the three countries are all implementing knowledge-based plans, they are not as comprehensive as either that proposed in this report or that proposed by the Korean authorities. The core of the plan in these three countries has focused primarily on the national information infrastructure pillar. However, they are now expanding their focus to include R&D, innovation and education.

2. In the case of the knowledge-based development strategy, the Korean authorities have held two public hearings, one in October 1999, the other in March 2000.

3. The establishment of a higher education council could be considered as a means of providing guidance on how to adapt Korean colleges and universities to world-class levels. Such a council would be made up of all the important stakeholders in the education system – parents, the private sector, education specialists, professors and government – and would set guidelines and monitoring and evaluation procedures for the allocation and effective use of public resources. A separate review board could also be set up to examine quality and accountability issues, and to promote more transparent and timely information on the performance of graduates from different types of public and private education facilities.

4. Increasingly, policy issues for lifelong learning go beyond the traditional confines of education ministries and involve others, including the Ministry of Labor and the Ministry of Information and Communications. In some areas, such as the Internet, it is not even clear where the major responsibility lies. So, it is imperative that policy development is done in consultation with the relevant agencies in government and between the government and segments of the private sector and civil society that have a direct bearing on the sectoral strategies being formulated.

5. We say this based on our experiences in other countries. This is supported by a recent survey of 200 key opinion makers in government, the private sector and civil society in Korea. The survey examined Korea's preparedness for the KBE. The main recommendation of the survey, and one that we fully endorse, is that it is not useful to discuss a plan for making Korea into a knowledge-based society in the absence of a specific implementation plan. Such a plan must include not only what will be implemented, but how and why it will be implemented and by whom. It should also include how the degree of success of the implementation will be measured and evaluated (survey by *Maeil Business Newspaper*, September 1999). In a number of countries, good plans and programs were wasted for they were not well implemented. The opportunity cost of such a failure would be extremely high for Korea given the mounting pressures for shifting to higher productivity-driven growth.

6. Finland, Ireland and other developed countries have recently embarked on "flagship" projects. These are considered to have large spillover effects for the acquisition, creation, dissemination and use of knowledge by society.

7. Many measures target the development of SMEs, high-technology start-ups, new venture firms, etc. They are administered by various agencies (MOCIE, MOST, MIC, Ministry of Culture and Tourism, etc.). In place of the current focus on providing financial and other support through institutions, specific services targeted to the needs of various enterprises may be warranted. Despite the measures, the bulk of SMEs are generally weak in terms of technological competitiveness (see Chapter 6).

8. The Spearhead Network Service disseminates knowledge and expertise on information and transfers among the spearhead projects. At the same time, this Internet service also serves those interested in the development of the Finnish information society by providing them with information on the progress of the Spearhead Projects (SITRA, 1998).

9. For the year 2000 budget, major plans are: to increase R&D investment from 3.7% of the national budget in 1999 to 4.1% in 2000, and to 5% by 2002. The plan for R&D in 2000 is to spend KRW 200 billion on promoting graduate level research in line with *Brain Korea* 21. Other plans include: increasing access nodes to the high-speed telecommunications network from the current 107 to 144; raising spending for information technology in the government from KRW 537 billion to KRW 671 billion; providing tuition support for kindergarten and secondary education pupils to cover 230 000 and 400 000 pupils, respectively; introducing college tuition loans to cover 300 000 students; allocating KRW 50 billion annually to the Cultural Industry Promotion Fund; increasing spending aimed at fostering the SME sector by 16% to KRW 2 294 billion; earmarking KRW 120 billion for supporting regional clusters, *e.g.* shoe manufacturing in Pusan (Ministry of Planning and Budget, 1999). The rationale for these specific goals needs to be discussed and explained; and evaluations should be carried out of the

outcomes, impacts and progress made in attaining these goals so that the programs can be adjusted and maximum impact obtained at least cost.

10. Both Finland and Ireland have also carried out surveys on the impacts of the general strategy and specific initiatives. See, for instance, SITRA (1999) and Information Society Commission (1999).

11. See the New Zealand exercise on this, which included the following: establishing the context; setting up the strategies; decisions made; and implementation of new strategies (Christie, 2000).

References

Christie, Rick (2000),
   "Steps towards a Knowledge Economy: The New Zealand Experience", paper presented at the "International Brainstorming on the K-Economy", organized by ISIS Malaysia, 27-28 May, Kuala Lumpur.

Cole, George (1999),
   "Great Leap Forward", *Times Educational Supplement*, 15 October.

Information Society Commission (1999),
   Ireland's Progress as an Information Society, Ireland, October.

Irish Inter-Departmental Implementation Group (1999),
   "Progress on Implementing the Information Society: Second Report of the Inter-Departmental Implementation Group", July, Ireland.

Korea Institute for Industrial Economics and Trade (2000),
   "Theory and Practice of the Knowledge-based Economy", Seoul: KIET.

Ministry of Planning and Budget (Korea) (1999),
   The Budget of the Republic of Korea: Fiscal Year 2000, December.

SITRA (1998),
   "Quality of Life, Knowledge and Competitiveness: Premises and Objectives for the Strategic Development of the Finnish Information Society", Helsinki.

SITRA (1999),
   Challenges of ICT in Finnish Education, Helsinki.

Woo, Cheonsik (2000),
   "The Road to KBE: The Case of Korea", paper presented at the "International Brainstorming on the K-Economy", organized by ISIS Malaysia, 27-28 May, Kuala Lumpur.

Annex I

Preliminary Assessment of the Knowledge-based Economy in Korea

The World Bank has developed a methodology to rank economies in terms of a set of 200 indicators that serve as proxies for the four areas identified in Section C of the Executive Summary. This methodology consists of a set of structural and qualitative variables that serve to benchmark how an economy compares to its neighbors or competitors or to countries it may want to emulate.[1] It helps to identify where a country may be weak and where, depending on its specific conditions and its development strategy, it may need to focus policy attention or future investments. This is the first step in a more detailed assessment based on in-depth work, analysis and interviews in view of developing a coherent national strategy for the more effective use of knowledge for development.

As a large set of variables is unwieldy, we have developed a simplified "knowledge assessment scorecard" consisting of 20 variables that attempt to capture the essence of a country's preparedness for the KBE. Each of the variables has been normalized on a scale of 0 to 10, so that the highest value is rated 10 and the lowest, 0. The comparison is undertaken for a group of 60 economies which includes most of the developed economies and about 30 developing economies. An economy should not necessarily aim for a score of 10 on all variables. Some reflect performance; others reflect trade-offs, as will be seen below, which characterize different development strategies. Still others reflect the particular structural characteristics of an economy. The normalized variables are put on star diagrams to graphically illustrate and facilitate comparisons among countries. The scorecard for Korea is presented below in Figure 1a.

Two indicators are used to illustrate the overall performance of a country: *annual GDP growth* 1990-97, and the *human development index* 1997. The human development index is a composite measure of three components: longevity (measured by life expectancy); knowledge (adult literacy rate and mean years of schooling); and standard of living (real GDP per capita in purchasing power parity). On these two measures, Korea ranks relatively close to the top performers.

For proxies of the economic regime, we use two indicators. For an index of the robustness of the financial system and country risk, we use the *composite ICRG risk rating*. To measure the degree of competition, we use *tariff and non-tariff barriers*, which is a composite of the rating on tariff and non-tariff barriers and customs corruption from the Heritage Foundation and the *Wall Street Journal's* economic freedom rankings. On both these measures, Korea rates just above and at the mid range respectively, suggesting that the country still has some way to go to figure among the top-ranking countries.

For the institutional regime, we use four measures. As a proxy for government quality, we use the *corruption perception index*, since the government plays a key role in setting the institutional and regulatory regime. On this measure, Korea ranks in the bottom third. For the general legal regime, we use compliance with court rulings. Because adequate protection of intellectual property rights is a critical issue in the incentive regime, we also use businessmen's assessment of *intellectual property protection*. On both these legal/institutional issues, Korea ranks only in the middle. Finally, as a measure of the free exchange of ideas, we use *freedom of the press*, where it ranks roughly at 75%. Taken together, these measures suggest that Korea still lags behind on the institutional side. Chapter 2 provides a broader and more detailed analysis of some of the critical issues that need to be addressed in this area.

For education and human resources, we use three variables: the *literacy rate* gives a very broad stock measure, while *secondary and tertiary enrolment rates* provide a flow rate. In terms of literacy, Korea ranks very near the top, and on the enrolment rates, it ranks in the top 20%. These measures reflect the very impressive gains the country has made in educational attainment. However, on the third variable, which entails a qualitative ranking of the *flexibility of people*, Korean is placed in the bottom third. This foreshadows a point which is developed in Chapter 3, *i.e.* despite the country's very high educational attainments, there is significant concern among the global business community about the quality of Korean education and the creativity of Korean students.

For the innovation system, six variables are used as this area is of special relevance to a knowledge-based economy. As proxies for the extent to which an economy taps into the growing stock of global knowledge, we use *foreign direct investment as a share of* GDP and *royalty and licensing fees paid abroad as a percentage of* GDP. On both these measures Korea ranks at the bottom, suggesting that it is not well integrated into the global knowledge system. Two measures are used as indicators for R&D. For effort we use *scientists and engineers in R&D per million population*, while for domestic R&D output we use *patents granted in the United States to different countries per million of their population.*[2] On scientists and

Annex Figure 1a.   **Knowledge assessment scorecard: Korea**

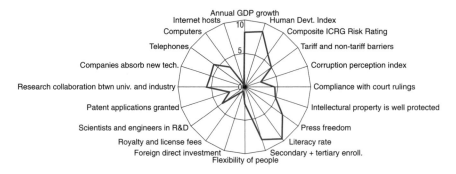

Annex Figure 1b.   **Knowledge assessment scorecard: United States**

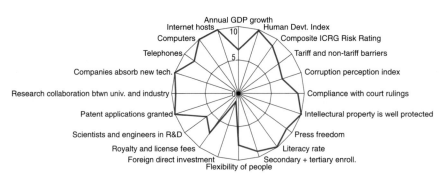

Annex Figure 1c.   **Knowledge assessment scorecard: Singapore**

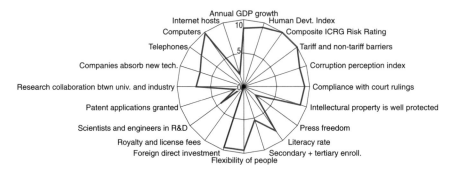

Annex Figure 1d.   **Knowledge assessment scorecard: Japan**

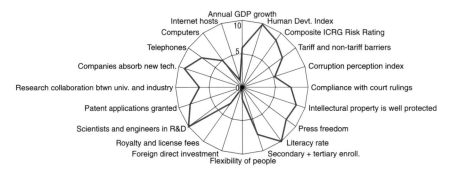

engineers, Korea ranks below the middle range, while on patenting it ranks in the bottom third – which should be a matter of concern for a country that aims to become a knowledge-based economy. Finally, to obtain an idea of the actions of some of the critical actors in the innovation system, we use qualitative rankings about the *closeness of research collaboration between universities and industry* and *how well companies absorb new technology.* On these indicators, Korea ranks in the middle range, suggesting that in this area too it has a significant way to go. These issues are analyzed extensively in Chapter 5.

Three variables are used to describe the information infrastructure. *Telephones per 1 000 population* is the sum of telephone mainlines and mobile phones and provides a better indicator of connectivity than either in isolation. *Computers per 1 000 population* is an indicator of personal computer penetration. *Internet hosts per 10 000 population* measures use of computers and communications. On telephones, Korea does slightly better than mid-range; on computers, it ranks lower than mid-range and on Internet hosts, it lags seriously. However, the use of Internet has exploded recently, so the current situation is more favorable than suggested by the data. Information infrastructure is analyzed in more detail in Chapter 4.

To provide some perspective on how Korea compares to other economies, Annex Figures 1b to 1d present the knowledge assessment scorecards for the United States, Singapore and Japan. These countries have been chosen as illustrations of different strategies. The United States, which is arguably the most advanced knowledge-based economy, shows a relatively full circle. However, it is not fully open to international competition as can be seen from its ranking on the tariff and non-tariff indicator. Also, it does not rely very much on direct foreign investment, in large part because it has such a strong domestic R&D infrastructure.

Singapore, on the other hand, is very open to the world and obtains the majority of its technology by relying very heavily on foreign investment and technology transfer. It is not so strong in its R&D efforts (although these include R&D undertaken by multinationals located in Singapore), and is very weak in terms of patenting. However, it does very well on the extent to which companies absorb foreign technology and on research collaboration between university and industry. Singapore does not do well on freedom of the press, reflecting the very strong role of the state in economic and social activities. On the other hand, it does quite well on all the indicators of the information infrastructure and is one of the economies that has most systematically used ICTs as part of its development strategy.

Japan looks like a more advanced version of Korea, which is not surprising since Korea has followed the Japanese development model in many respects. However, given the decade-long slump that Japan has been experiencing, Korea has begun to look at the way in which other economies, particularly the United States and some of the dynamic European countries, are structuring their economies to make more effective use of knowledge. All these issues are analysed in the present report.

Notes

1. The quantitative variables are based on official statistics compiled by the respective country governments and vetted by international statistical organisations. Most are reported in the World Bank's *World Development Indicators* 1999. The qualitative indicators come primarily from surveys of businessmen carried out annually by the IMD and the World Economic Forum, although some are provided by specialized organisations such as Transparency International or the Heritage Foundation. The sources for each of the variables on the scorecard are provided at the end of the annex.

2. Because countries have different patenting procedures, rather than using patenting in their home countries, we have used patenting in the United States, which is the largest market for patenting. This biases the ranking in favor of the United States, but in any case the United States is recognized as the world leader in innovation, and the measure does give a good indication of the relative performance of other countries. Japan, for example, scores very close to the United States.

Knowledge Assessment Scorecard Indicators

Performance Indicators

1. Average annual GDP growth 1990-97 (%) (World Development Indicators, 1999)
2. Human development index 1997 (Human Development Report 1999, UNDP)

Economic Incentives

3. Composite ICRG risk rating (December 1998) (World Development Indicators, 1999). This is the Composite International Country Risk Guide: an overall index, ranging from 0-100, based on 22 components of risk.
4. Tariff and non-tariff barriers (Heritage Foundation, 1999)

Institutional Regime

5. Corruption perception index (Transparency International, 1999)
6. Compliance with court rulings (World Economic Forum, 1999)
7. Intellectual property is well protected (World Economic Forum, 1999)
8. Press freedom 1999 (Freedom House, 1999)

Human Resources

9. Literacy rate as % of population (IMD, 1999)
10. Secondary + tertiary enrolment, 1996 (World Development Indicators, 1999)
11. People are flexible to adapt to new challenges (IMD, 1999)

Innovation System

12. FDI as % of GDP 1997 (World Development Indicators, 1999)
13. Royalty and license fees as % of GDP, 1997 (World Development Indicators, 1999)
14. Scientists and engineers in R&D per million population, 1985-95 (World Development Indicators, 1999)
15. Patent applications granted in the US to different countries per million of their population (US Patent and Trademark Office, 1998)
16. Research collaboration is very close between universities and industry (World Economic Forum, 1999)
17. Companies are aggressive in absorbing new technology (World Economic Forum, 1999)

Information Infrastructure

18. Telephone per 1 000 persons, 1997 (telephone mainlines + mobile phones) (World Development Indicators, 1999)
19. Computers per 1 000 persons, 1997 (World Development Indicators, 1999)
20. Internet hosts per 10 000 persons, July 1998 (World Development Indicators, 1999)

Closing the Digital Divide: Some Country Examples

With the growth of the Internet revolution, an emerging policy concern is the growing "digital divide" within and among countries. In general, this new form of inequality stems from disadvantaging factors such as geographic location, gender, income and education. Rural communities, women, the poor and the illiterate are most likely to be excluded from the opportunities to learn and adapt to new technologies. A number of governments have developed strategies and programs aimed at closing the divide; their experiences are described below.

Country	Strategies	Programs/Projects
Denmark	• Promoting gender equality in IT use, research, education	• The universities were asked to prepare a "Plan of Action for Equality" to improve female enrollments and to promote a more equal distribution of the sexes among scientific personnel. • In addition, the universities were encouraged to recruit more women students and researchers in research and IT-based study programs. • Special training programs have been introduced, targeting unemployed and unskilled female workers. • A new training program is being developed in close co-operation with the wider social community.
Ireland	• Promote general access to ICT throughout the community. • Promote "access" as awareness, physical accessibility, usability and user-friendliness of information technology and availability of tuition and technical support. • Promote training for the general public, targeting disadvantaged and marginalized groups.	*Library projects:* • Maximize the number of public Internet access points in libraries by escalating the investment. Libraries must extend opening hours and provide training programs for the general public and staff. Further initiatives to promote accessibility of IT equipment for the disabled. *School projects:* • Schools should act as central facilitators of community access to ICT. Upon receiving government IT funding, schools should provide greater community use of their IT equipment. IT 2000 program targets at least 50% of schools with IT access. *Community and voluntary sector:* • The Department of Social, Community and Family Affairs will rapidly increase the level of funding to the community and voluntary sectors. • The Information Society Commission is establishing focus groups on Connected Communities, especially to overcome the technical barriers to late adopters and people with special needs. • With the co-operation of the private sector, the Ennis Information-age Town project resulted in a total of 49 towns throughout the country developing strategies for improved the use of ICT in their own communities. *Citizen Information Centers:* • 28 Citizen Information Centers and additional outreach locations around the country provide access to a Citizens' Information Database for the provision of information on social services and job search. *Raising awareness of ICT:* • The Information Society Commission has invited the Internet Service Providers (ISP) to develop simple, easy-to-use Web-based e-mail services directed to late adopters.

Country	Strategies	Programs/Projects
Canada	• Develop a national access strategy to ensure affordable access by all Canadians to essential communication services. • Emphasize convenient, less expensive and just-in-time lifelong learning to improve the ability of the labor force.	*National access*: • The federal government has established a working group on access and is undertaking a study of barriers to access to the Information Highway, focusing on the influence of social factors (age, disability, ethnicity, education and income). • Government has increased the budget of the Community Access Program to expand the number targeted by the Federal Youth Employee Strategy. *Community access*: • Industry Canada's Community Access program enables people in rural and remote communities to access the Internet. Electronic government service and information is being delivered to these communities with help to enable them to develop skills. • The SchoolNet and LibraryNet programs facilitate the connectivity of Canadians by equipment donation from government, corporations and individuals. In 2000, Canada will establish up to 10 000 public access sites to connect all Canadians. *Preferential tariffs for education and health entities*: • In September 1996, Canada allowed preferential tariffs on telecommunications services for non-profit educational and health service bodies. These tariffs apply only to services provided on a competitive basis and must cover the costs of providing the service.
Singapore	• Extend the use of PCs and Internet to low-income households. • Address language and attitude issues in addition to income and gender issues.	• Provide a free used PC and basic training to 30 000 low-income households. • Provide free broadband Internet access at community centers and locally relevant Web content in other Asian languages with the help of the private sector and tax incentives. • Develop national Chinese Internet programs for different population segments to bridge the language gap. • Extend the "e-Ambassador" program, *i.e.* use current Internet users to voluntarily teach latecomers. The program will be extended to around 2 500 new e-ambassadors.
United States	• Lower the prices of hardware and software to make them affordable to consumers by adapting pro-competition policies. • Provide tax incentives to encourage the private sector to donate computers to communities and to stimulate training and education programs. • Improve information infrastructure in rural areas. • Encourage women to pursue technical education and a technical career.	• Universal Service Program: In 1998, the Federal Communication Commission's (FCC's) Universal Service Fund was established to help low-income households pay monthly service bills and the installation costs required to initiate services. • The US Department of Agriculture's rural service provides lending and technical consulting in rural areas. • Using female role models in the media to improve the image of women scientists and technologies. • Developing science encouragement programs to increase female students' interest in technical careers. • Using influential colleagues, mentors and sponsors in schools and business communities to develop women's interest in IT. • Using the alliance between business, government and education communities to expand technical training and job opportunities for women.
United Kingdom	• Develop strong regional partnerships and strengthen the roles of the regional development agencies (RDAs). • Increase communities' access with small enterprise strategies.	• New regional venture capital funds will specialize in providing small-scale equity to businesses with growth potential, drawing on local enterprise. • Business-led RDAs will act as lead bodies at the regional level for co-ordinating inward investment, raising skills, encouraging links with business and with higher education. • The government works with business in the communities to extend "IT for All" to the "information have-nots" • All public libraries will be linked electronically by 2002 to provide community access to digital technology.

References

Information Society Commission (2000),
   IT *Access for All: Report of the Information Society Commission, http//:www.isc.ie,* Ireland, March.

Johnston, David (2000),
   "Canada: Life-long Learning in the Knowledge Based Economy", paper prepared for the "Second Global Knowledge Conference", Malaysia, March.

Kim, Eun Jeong (2000),
   "Summary of Strategies for the Digital Divide in a Knowledge-based Economy: Denmark, Ireland, Singapore, and USA", World Bank.

UK Department of Trade and Industry (1998),
   Our Competitive Future: Building the Knowledge-driven Economy, December.

Annex III

Some Reflections on the Challenges of Policy Implementation

There is considerable information available to governments on what needs to be done to strengthen the basis for the development and use of knowledge. More tricky is the actual implementation of the action programs. This typically includes overcoming political resistance and reducing the risk of unmotivated deviations in policy. A policy which looks good on paper can in fact have detrimental effects if market actors anticipate that it will soon be overturned.

For policy to be consistent and credible, broad support within (and outside) government for long-term objectives is key, not least as it can help to underpin the long-term commitment to these objectives. Policy packages which span a number of areas can serve to strengthen political backing by increasing the number of winners and weakening the position of groups that will have to give up privileges. Improving policy co-ordination, however, often requires co-ordination and decision making which involve different policy areas and cross the traditional delineations of administrative competence. This often requires a new institutional set up.

Appropriate incentive systems are needed to bring about policy co-ordination. Financial pressures can be used creatively to spur change in governance and to encourage the adoption of assessment mechanisms designed to induce innovative behavior. Checks must be put in place against government failure, such as institutions furthering their own special interests and adopting a partial rather than an economy-wide perspective.

In Finland, the Netherlands and Norway, various concerned ministries set themselves the common goal of developing the information society, demonstrating the progress that can be made once joint objectives have been identified. On the other hand, experience also shows that rapid progress may require sanctioning by the highest level of authority. The knowledge-based economy relies partly on providing appropriate conditions for initiative and creativity from the bottom-up, and the need for the top to embrace this principle is particularly important in relatively autocratic societies.

How can policies in individual areas be integrated into a broader package developed in consultation with the social partners to ease transition problems? One strategy is to begin with those measures which appear to be the most feasible, universally supported and whose effects are likely to be the most evident. Once these measures have been in existence for some time and their effects have been evaluated, necessary corrections can be implemented and more difficult decisions can be pushed through. Policies conducive to improved knowledge creation and use in Finland, Iceland, Japan and the Netherlands have been able to evolve along these lines. Even when a "big-bang" strategy has been advocated, some key knowledge-related policies (technology, education, etc.) have generally evolved gradually over a period of decades (*e.g.* New Zealand). On the other hand, the ability to advance may hinge on the political will to push through difficult decisions, handle the associated transition costs and demonstrate positive outcomes. In some countries, a crisis situation has helped muster support for reform (*e.g.* Finland, Japan, Korea). It is important that policy makers exploit such opportunities as they arise.

Many governments in the OECD are increasingly making use of other instruments to facilitate or strengthen policy implementation. These include:

- Decentralisation of responsibilities can help to bring policies closer to their constituents and can allow for adaptation and specialisation, *e.g.* in education policy or mechanisms for establishing science-university linkages. On the other hand, in areas where wider regional and social concerns need to be taken into account, decisions must not be left to the local level as this would result in compartmentalisation and bring about new problems of consistency and co-ordination.

- Private-public partnership can be very important for making policies more responsive to customer needs and to strengthen political support.

- Large-scale information campaigns can help to anchor policies more broadly in society and enable policy design and implementation which respond to real needs.

- "Audits" and international benchmarking, *e.g.* of how policy organisation and formulation relate to economic behavior and performance, can help to raise general awareness and also induce a process of self-examination in governments.

- Large-scale initiatives can be undertaken to involve society in policy formulation and the establishment of new priorities in science and innovation. For instance, both the United States and Australia have recently held major summits on innovation policy. The US National Science and Technology Council hosted a Summit on Innovation at the end of 1999, aiming to explore the future direction and properties for federal support of innovation. The summit brought together business, government, the research community and non-profit organisations to examine both obstacles to and opportunities for greater innovation.

Regarding important linkages between policy areas, measures that promote broad-based upskilling and lifelong learning can help to raise the mobility and employability of workers and mitigate the costs of job displacement. Social security programs and transfers protecting social cohesion will continue to play a key role in preserving a social fabric conducive to trust, itself a major building block for risk-taking, innovation and creativity in a broader sense. At the same time, it is crucial that policies be designed in such a way that they do not undermine incentives for work, upskilling, organisational change or restructuring. The OECD countries face a major challenge in putting into place, and successfully communicating to the general public, a comprehensive range of policies which set the frame for a mutual strengthening of social cohesion, on the one hand, and technological progress and change, on the other.

There can be a mutually reinforcing interplay between reforms and economic performance. In recent years, regulatory reform and intensified competition have been powerful engines of change in many countries, enabling strong performance improvements in sectors such as electricity, gas and water, transport and communication, wholesale and retail trade, finance and many other services. In some sectors, notably telecommunications and electricity, technological change was the main instigator of regulatory reform and has resulted in the erosion of the natural monopoly character of these industries.